PRAISE FOR

The Social Structure of High-Energy Physics Experiments at Fermilab

Mark Bodnarczuk's book is truly unique and will resonate with those interested in studying the history of big fundamental science. In the objective search for the most fundamental scientific truths, it highlights the subjective influences that human nature exerts on the practice of science. Bodnarczuk documents and codifies the changing dynamic of high-energy physics research from the post–WWII period to the present. His observations and conclusions appear universal, providing a predictive model for the successful conduct of research on the history of high-energy physics experiments going forward.

Bodnarczuk describes how the social structure of high-energy physics experiments at Fermilab was shaped and defined by the increasing scale, cost, and complexity of detectors used to explore physics at ever higher energies. He successfully applies his analysis to fixed-target and large collider experiments like ATLAS at CERN, but the fifty-year history of high-energy physics research at the SLAC National Accelerator Laboratory is beyond his scope.

At SLAC, I was a part of two Nobel Prize–winning experiments with the Mark I detector; one in 1974 for the J/Psi discovery and another in 1976 for the tau-lepton discovery. This was followed by the detailed studies of charm meson decay in 1978 with the Mark II detector, and its later discoveries at higher energy colliders, PEP-1 and SLC. So, as a laboratory that used electrons not protons to probe nature, SLAC provides an independent test of the author's conjectures on the technical and sociological drivers of particle physics experiments.

As I read this book, I recognized the evolution and drivers that the author describes, and that the concept of experiment "strings" was once again demonstrated by the Mark I, Mark II, and Mark III experiments I was a part of. Indeed, the evolution of SLAC confirms that Bodnarczuk's observations are applicable in a very broad fashion. They are indeed a checklist that leads to the success of each new enterprise, as high-energy physics research continues to be highly driven by close variants of the five resource economies the author describes—experimental real estate, beamtime, computing, the physics, and the physicist manpower.

—**RAFE H. SCHINDLER**, Professor of Particle Physics and Astrophysics, Stanford University, SLAC National Accelerator Laboratory

Mark Bodnarczuk's book takes the reader into the scientific, social, and interpersonal world of high-energy physicists and their experiments at Fermilab in the 1970s, 1980s, and 1990s with a window into these important and challenging times. The *Social Structure of High-Energy Physics Detectors at Fermilab* will provide valuable insights to scientists and students of science alike about how fixed-target experiments began and ended in the era before first generation collider experiments like CDF and D0, and today's massive ATLAS and CMS detectors became the primary method for doing high-energy physics.

—**JOHN P. CUMALAT**, Professor of Distinction,
Department of Physics, University of Colorado Boulder

Bodnarczuk describes in detail the sociological factors that govern research collaborations and publications in the world of high-energy physics and one might be tempted to think that this field is unique in terms of the forces that form it. My observation from a different vantage point—that of a materials scientist who has used neutrons and X-rays to probe the structure and dynamics of materials on which many of our modern technologies depend—is that many of the shaping forces are the same even though the currencies that Bodnarczuk describes may be different. Access to beam time and experiment strings are common currencies for neutron and X-ray experiments whereas computing power is generally less constrained because the level of sophistication needed has not reached those required by our high-energy colleagues. On the other hand, collaborations with sample growers, users of complementary techniques, and theorists to explain the data become important for success in materials research. While the detailed currencies change, the same general social principles described by Bodnarczuk broadly apply. For this reason, his analysis will be of interest to anyone trying to make sense of government-funded scientific research in the post–WWII era, a time during which the spin-offs from what often started as curiosity-driven research have increasingly defined modern life. There is much to learn here about how humans have organized themselves to make such remarkable progress in such a short period.

—**ROGER PYNN**, Emeritus Professor of Physics,
Indiana University, Bloomington

Mark Bodnarczuk's book studies how large experiments and their follow-up experiments form long-lived research traditions within the experimental program of a large high-energy physics laboratory. He calls these traditions "experiment strings" and explains that their formation is a defining feature of the "bigger big science" phenomenon that many historians now refer to as "megascience." Bodnarczuk also describes how over time the social structure of the experimental working groups—including the physicists, postdocs, graduate students, engineers, and technical support staff—begins to match the structure of the detectors' own subsystems, creating a "people-spectrometer system." He shows how social factors are inextricably involved in every phase of the experimental process, from the initial detector design through the final publication of results and suggests that the experiment strings at a laboratory can be seen as invisible mini-institutions corresponding to particular experimental programs, such as the study of charmed particles described in this book. At the time Bodnarczuk published his initial historical probing of the experimental strings at Fermilab using his physics and detector-driven analysis, he revealed new insights into the phenomenon of megascience, showing, for example, how it was shaped by the restricted funding for physics at Fermilab after about 1970. In today's world of high-energy physics, this approach to studying the historical development and social structure of the current generation of gigantic collider detectors (e.g., ATLAS and CMS within the context of laboratory life at CERN) offers a promising path for historians of science to explore.

—**LILLIAN HODDESON**, Professor of History Emerita and Thomas Siebel Chair in the History of Science Emerita at the University of Illinois at Urbana–Champaign

In *The Social Structure of High-Energy Physics Experiments at Fermilab*, Mark Bodnarczuk presents a unique and compelling examination of how scientific discovery emerges not only from physics theories and experimental data, but also from the social, organizational, and technological structures that make large-scale science possible. Drawing upon his firsthand experience working at Fermilab during the formative years of modern particle physics, Bodnarczuk offers a rare insider's perspective on the evolution of high-energy physics experiments and the human systems that supported them.

The book's central thesis is that the development of particle physics experiments is shaped by a combination of scientific objectives and what the author calls "resource economies," including experimental real estate, beam time, computing resources, and scientific manpower. These factors influence the formation and evolution of research collaborations, detector technologies, and experimental programs. Bodnarczuk introduces the concept of "experiment strings," showing how one experiment builds upon the technical, organizational, and intellectual foundations of its predecessors. This framework provides a powerful lens for understanding the growth of increasingly complex scientific enterprises, from Fermilab's fixed-target experiments to modern collaborations such as ATLAS and CMS at CERN.

One of the book's greatest strengths is its interdisciplinary approach. Rather than focusing solely on physics results, the author examines science as both "knowledge" and "practice." He argues that scientific publications represent only the outcome of a much larger process involving engineers, technicians, operators, managers, and support staff. This perspective highlights the often-overlooked contributions of the broader workforce that enables scientific discovery. For readers involved in large scientific facilities, engineering organizations, or national laboratories, this insight is particularly valuable because it mirrors the realities of modern megascience projects.

The historical narratives and personal recollections woven throughout the text add depth and authenticity. Bodnarczuk's descriptions of operating Fermilab beamlines, working in the Proton Area, and observing the evolution of detector technologies provide vivid examples of how science is conducted in practice. His reflections connect technical development, organizational behavior, and human ambition into a coherent story of scientific progress.

While the book is rooted in high-energy physics, its broader lessons extend well beyond particle science. Researchers, engineers, project managers, and leaders of large technical organizations will recognize familiar themes involving resource allocation, collaboration, innovation, and organizational complexity. The author successfully demonstrates that scientific breakthroughs are as much social achievements as technical ones.

Overall, *The Social Structure of High-Energy Physics Experiments at Fermilab* is an insightful and thought-provoking contribution to the history and sociology of science. It offers a fresh perspective on how major scientific discoveries are produced and provides a valuable framework for understanding the organizational dynamics behind some of humanity's most ambitious research endeavors. For anyone interested in the intersection of science, engineering, and organizational systems, this book is highly recommended.

—**MOISES SMART, PHD**, Engineering Division Director at Argonne National Laboratory's Advanced Photon Source (APS)

I read *The Social Structure of High-Energy Physics Experiments at Fermilab* by Mark Bodnarczuk as an outsider from the inside. From the outside, because I am not a high energy physicist. From the inside, because I have spent my professional career at the SLAC National Accelerator Laboratory, the electron accelerator laboratory complement to the proton accelerators as Fermilab. What makes the story most compelling reflects this duality.

I look at the scientific success captured in *The Social Structure* and the trajectory for science this structure established. I also see how the structure of these prior successes now box in the high-energy physics community's process for science with projects and experiments of ever-increasing scope, scale, and duration that inhibit dynamism and threaten the creativity and ingenuity foundational to the success of high-energy physics in the second half of the twentieth century.

More importantly, I look at how team science and coordinated facility planning and investment has shaped my career in the science of *stuff*—chemistry, biology, and materials science. I wonder how we continue to pursue the boundary of what is possible, without being too slow with our execution due to cost and complexity. At the heart of science is constrained creativity bounded by empirical observation. Without truly new observation, we cannot fuel the engine of scientific discovery. But projects that transcend the generational scale of a human life put the scientific endeavor at risk.

The Social Structure of High-Energy Physics Experiments at Fermilab helps explain how high-energy physics has arrived at the present and how that trajectory should inform the future of discovery-driven science. The concept of an experimental string provides a valuable framework for planning the future because it highlights how prior investments create a mortgage for future innovation. We must learn how to right-size the risks associated with discovery-driven science to avoid sacrificing creativity in favor of security.

—**KELLY GAFFNEY**, Professor of Photon Science, SLAC National Accelerator Laboratory and Stanford University

The Social Structure of High-Energy Physics Experiments at Fermilab

A Physics and Detector Driven Analysis

Mark Bodnarczuk

BRECKENRIDGE
PRESS

Cover design by GKS Creative

Text design by GKS Creative

ISBN: 978-1-963461-23-7

Library of Congress Control Number: 2026900567

Subjects: LCSH: 1. Standard model (Nuclear physics) – History. 2. Particles (Nuclear physic) – History.

3. High-energy physics (Experimental) – History.

Published by

The Breckenridge Press

BRECKENRIDGE
PRESS

This book is dedicated to

THOMAS LARSON BODNARCZUK

MAY 16, 2002–MAY 2, 2021

In death, as in life, you continue
to have an enormous impact on us.

ALSO BY MARK BODNARCZUK

Diving In: Discovering Who You Are in the Second Half of Life

Island of Excellence: 3 Powerful Strategies for Building a Culture of Creativity

The Breckenridge Enneagram: A Guide to Personal and Professional Growth

Making Invisible Bureaucracy Visible: A Guide to Assessing and Changing Organizational Culture

Finding New Life After the Death of My Son

"There is much in common between what the creative artist does and what the scientist does . . . They both usually pass through a morose period when their contemplation is deep, and mostly at a subconscious level. Finally, what they do create just comes bubbling up from the subconscious to the conscious mind. Inevitably, this inspirational act is accompanied by a period of euphoria, an ecstasy which has been documented time and again for scientist and artist alike."

Robert R. Wilson
The Humanness of Physics

TABLE OF CONTENTS

Foreword

I was a graduate student in the Tagged Photon Laboratory (TPL) on experiment E-25 in the Proton Area at Fermilab and I helped install the TPL beamline described in this book. I became a postdoc and the first Robert Rathbun Wilson Fellow at Fermilab, then faculty member at the University of Colorado Boulder and helped design and perform a series of charm particle production and decay experiments i.e., experiment E87A, E401, and E400 in the Proton East enclosure EE-4. So, I was one of the physicists on an experiment string similar to the one described in this book. Mark Bodnarczuk was one of the Operations Crew Chiefs who ran 400 GeV protons to the primary production targets for my experiments.

I subsequently helped to design and perform experiments E687 and E831 in the Proton East Wideband Lab to do further studies of charm particles, with a strong overlap to the physics explored in the Tagged Photon Laboratory case study of experiments E516 and E691 described in what follows. As the primary beam energy for the Wideband Lab string of experiments rose to 800 GeV, the size and complexity of the detectors, the number of channels of electronics, and the number of collaborators grew in ways described by Bodnarczuk's physics and detector driven analysis.

The Tagged Photon and Wideband Lab experiment strings are excellent examples of how science and institutions evolve over time in terms of workforce development. Scientists designed and developed detectors at home institutions and involved undergraduate and PhD students. The equipment was tested and installed at Fermilab, everyone ran data acquisition shifts to gain exposure working in a large collaboration, and then everyone was involved in the physics analysis that led to final publications.

Strings of experiments conducted in the Tagged Photon and Wideband Labs are also excellent examples of how the search for physics at higher energies required the development of the first trigger processors in high energy physics, along with high-resolution silicon microstrip vertex detectors used to identify and reconstruct the decay vertices of short-lived charm particles.

Despite trying sophisticated triggers, the most successful strategy was to accumulate as many events as possible with minimal requirements. Thus, the Tagged Photon and Wideband Lab experiment strings dramatically increased the online and offline data acquisition requirements, resulting in thousands of compact data tapes. For example, experiment E691 in the Tagged Photon Lab and E687 in the Wideband Lab each accumulated 10,000 fully reconstructed charm particle events, with E687 going on to record over 500 million triggers and 100,000 fully reconstructed charm particle events during the 1990-1991 run.

It was no longer feasible to reconstruct the events in a single mainframe at Fermilab so multiarray processors were developed with the added bonus of significant cost savings. The Advanced Computer Program (ACP) at Fermilab led the effort to develop the type of massively parallel systems that high-energy physicists now use to do particle physics data analysis. As Bodnarczuk mentions in chapter 2, this crucial advance in low-cost computing power originated from members of the Tagged Photon Lab experiment string led by Tom Nash.

I concur that the social factors involved in the practice of high-energy physics at large laboratories like Fermilab cannot be separated from the detectors, detector innovations (triggering), and computing framework (online and offline). These social factors are even more critical in the current era of collider physics at CERN. Bodnarczuk's approach to studying experiments within a laboratory context of resource economies like proton economics, computing economics, and experimental real estate helps explain how these social factors evolved over time in the search for higher and higher energy physics.

When Fermilab experiment E687 ran in the late 1980s and early 1990s, I was one of about 80 physicists from universities in Bologna, Frascati, Milano, Colorado, Illinois, Northwestern, and Notre Dame, and the detector included a high-resolution silicon microstrip vertex detector with over 8,000 electronic readout channels. I am currently a member of the massive CMS collaboration at CERN's Large Hadron Collider, with over 5,500 particle physicists, engineers, technicians, and students from over 200 institutes across more than fifty countries. The CMS detector has 140 million separate readout channels with the vast majority being in the silicon tracker (pixels and strips), which is a more complex iteration of the vertex detectors used to precisely identify particle paths at the Tagged Photon and Broadband Labs.

In chapters 3 and 4, Bodnarczuk discusses how the complexity of the scientific measurements and the uncertainty in what to expect at extreme energies led to highly instrumented general-purpose experiments like Fermilab's CDF and DO collider detectors that were operational at the time. In the preface, Mark argues that a physics and detector driven analysis could be used to study the ATLAS and CMS experiments, as well as the IDEA, CLD, and ALLEGRO experiments proposed at CERN's Future Circular Collider that promises to be the largest scientific enterprises ever attempted.

For those who took part in early fixed target experiments at laboratories like Fermilab, SLAC, Brookhaven, Berkeley, DESY (Germany), and KEK (Japan), Bodnarczuk's book is a meaningful remembrance of important milestones and developments in the field of high-energy physics. For physicists and graduate students who joined the field after the early fixed target experiment days, it serves as an informative guide to how the field developed, and an approach to studying the history and social structures of today's mammoth experiments. For scientists and students of science who are not part of the discipline of high-energy physics, the book will help to demystify how high-energy physics experiments begin and then end with publications that contribute to the Standard Model of the universe.

John P. Cumalat.[1]

1 John P. Cumalat is a Professor of Distinction in the Department of Physics at the University of Colorado Boulder. He is a Fellow in both the American Physical Society and American Association for the Advancement of Science and an author of more than 1500 journal articles. He served as Chair for 20 years. John has broad research interests in experimental particle physics, including charm physics, physics that is currently beyond the Standard Model of particle physics, and developing cutting-edge detector technology and instrumentation that can survive in extreme radiation environments. He is a member of the Compact Muon Solenoid (CMS) experiment which is one of two large general-purpose particle physics detectors built in the Large Hadron Collider at CERN in Switzerland and France. The CMS experiment is one of the largest international scientific collaborations in history.

The Proton Pagoda at Fermilab is a 30-to-40-foot-tall, glass-enclosed, two-level steel structure built in 1976 that served as the primary, 360-degree control center for the Proton Experimental Area's underground beamlines. Designed by DUSAF Project Architect, Don Moll, it provides a panoramic view of the Fermilab site and features a distinctive central cylinder with a double-helix, DNA-style staircase. It functioned as the control room for the P-West, P-Center, and P-East experimental halls. As one of the Proton Area operations crew chiefs, Mark Bodnarczuk worked in the Pagoda's control room to ensure that the 400–800 GeV proton beams from the main accelerator were transported to the Proton Area's primary production targets, with the secondary and tertiary beams produced subsequently transported to experimental apparatus in the downstream enclosures of the P-West, P-Center, and P-East experimental halls, including the Tagged Photon Laboratory, where the Tagged Photon Magnetic Spectrometer for experiment string E-516-691-769-791 was located.

Preface

The story of how I came to write this book resembles the winding, serpentine path of a labyrinth. The journey has been punctuated by what Wolfgang Pauli and Carl Jung called acausal synchronicities. These are meaningful connections between the Outer Labyrinth of the destiny I was *creating* with my day-to-day choices, and the destiny that I *had* along the Inner Labyrinth of what Jung and Pauli called the Self—the nucleus of the human psyche.[2] As the historical development of quantum mechanics and the theoretical and experimental path that led to establishing the Standard Model of physics clearly shows, Pauli wasn't the only physicist to wax eloquent on the philosophical, psychological, and theological implications of the scientific knowledge he created and to ponder the larger *meaning* and *purpose* of science.

From the *physicists' perspective*, I'm thinking about Werner Heisenberg's inquiry into Aristotle's work as the basis for his analysis of "language and reality in modern physics"[3] as well as Albert Einstein's claim that the profounder sort of scientific minds all possess a "cosmic religious feeling" that emerges from

2 Wolfgang Pauli viewed Carl Jung's notion of synchronicity as a bridge between quantum physics and psychology. He collaborated with Jung for twenty-six years to develop an "acausal" principle that described the connections between mind and matter, in ways that echoed the mechanism of quantum entanglement. Pauli viewed synchronicity as an important expansion of scientific understanding that moved beyond a purely mechanical, causal, and reductionist view of the universe. See C. A. Meier (ed) *Atom and Archetype: The Pauli/Jung Letters 1932-1958* (Princeton, NJ: Princeton University Press, 2001), and David Lindorff and Markus Fierz, *Pauli and Jung: The Meeting of Two Great Minds* (Wheaton, IL: Quest Books, 2004).

3 See Werner Heisenberg, *Physics and Philosophy* (New York: Harper & Brothers Publishers, 1958), 167-186.

the "Old One"[4] as a universal, non-anthropomorphic, Spinozan-like god as a symbol of the orderly, lawful nature of the universe—a view that echoes the underlying metaphysics of Jung's psychology.[5] Then there is Neils Bohr's discussion on the connection between the complementarity principle and Eastern philosophy,[6] and more recently, Stephen Hawking's materialist metaphysic that claims, "There is no God. No one directs the universe."[7] I have described some preliminary thoughts on the larger meaning and purpose of science in chapter 6 of this book, and I have elsewhere included a brief summary of my perspective on the metaphysics that underlies the high-energy physics experiments performed at Fermilab, and the theoretical-experimental edifice that is called the Standard Model of the universe.[8]

---

4 For Einstein's discussion on the "cosmic religious feeling" see Albert Einstein: *The World As I See It* (New York: The Wisdom Library, 1949), 26-29. A discussion of the "Old One" was also included in his December 4, 1926 letter to Max Born regarding Einstein's objections to the probabilistic interpretation of quantum mechanics, "Quantum mechanics is certainly imposing. But an inner voice tells me that it is not yet the real thing. The theory says a lot, but does not really bring us any closer to the secret of the 'old one.' I, at any rate, am convinced that *He* is not playing dice," see Max Born and Albert Einstein, *The Born-Einstein Letters 1916-1955: Friendship, Politics and Physics in Uncertain Times* (New York: Macmillan, 2005), 88. Einstein describes his view of god as being near to those of Spinoza's monism in an April 26, 1947 letter to Professor Marvin Magalaner at the City College of New York, see Banesh Hoffmann, *Albert Einstein: Creator and Rebel* (New York: New American Library, 1986), 95.

5 While the difference between Jung's psychology and his underlying metaphysics is still debated in the Jungian community under the auspices of the mind-body problem, Jung's direct references to monism in his *Collected Works* and autobiography, his view of the biological basis of the psyche and dreams, plus other statements indicate that the underlying metaphysics that Jung's god-archetype and god-image cohered in echoed Spinoza's dual aspect monism. On this view we only perceive two of an infinite number of the god-archetype's attributes, i.e., mind (psyche) and matter (physical and biological), and these are two aspects of a single underlying "substance," i.e., god and nature. For the current discussion on this in the Jungian community, see Jon Mills (ed), *Psychoanalysis and the Mind-Body Problem* (New York: Routledge, 2022), especially Jon Mills comments on page 8 and Gary Clark's comments on page 10. For references to monism in Jung's work see C. G. Jung, *The Structure and Dynamics of the Psyche*, vol. 8 of *The Collected Works of C. G. Jung*, trans. R. F. C. Hull, 2nd ed. (Princeton: Princeton University Press, 1969), 215 (§418), and C. G. Jung, *Memories, Dreams, Reflections* (New York: Vintage Books, 1989), 310-311, and 324.

6 See Niels Bohr, *Atomic Physics and Human Knowledge* (New York: Dover Books on Physics, 2010), 31, and D. S. Kothari, "The Complementarity Principle and Eastern Philosophy" in A. P. French and P. J. Kennedy (eds), *Neils Bohr: A Centenary Volume* (Cambridge, MA: Harvard University Press, 1985), 325-331.

7 See Stephen Hawking, *Brief Answers to the Big Questions* (New York: Bantam Books, 2018), 20.

8 See Mark Bodnarczuk, *A Biblical Approach to Jung's Psychology: Experiencing the Self— the Image of God Within* (Research Gate, December 7, 2025), 1-29.

From the *philosophers' perspective*, the graveyard of the philosophy of science is littered with bodies of knowledge that are interesting but widely discarded. I'm thinking specifically about the views of Rudolph Carnap's logical positivism, with its focus on meaningful scientific discourse, and scientific propositions being verifiable, and Karl Popper's view that scientific theories can never be verified, but rather a proposition is "scientific" only when it can be falsified. Unlike most of the physicists described above, both tried to press science into the service of being the ultimate adjudicator of the veracity and meaning of all human knowledge. Neither one had ever worked in a science laboratory like Fermilab. Nor did they, or other philosophers and sociologists of science, have firsthand, laboratory-based experience with the distinction between science-as-knowledge (outcome) and science-as-practice (process).[9]

Echoing Nietzsche's admonition about the failure to ground knowledge in history,[10] published scientific papers (science-as-knowledge) are produced after-the-fact, downstream of the creation of physical effects in the experimental apparatus (science-as-practice), many of which do not occur in nature unattenuated.[11] So, it's the planning, managing, operating, documenting, and recording activities that go into the production of a scientific publication (science-as-practice), that create the outcome (science-as-knowledge), and together they constitute what is called the map making function and purpose of science.[12] That's what this book is about—the day-to-day

9 We see this same lack of firsthand, laboratory-based experience in other philosophers of science. For example, Bas van Fraassen, *The Scientific Image* (New York: Oxford University Press, 1980), Paul Feyerabend, *Against Method* (New York: Verso, 2010), Imre Lakatos, *The Methodology of Scientific Research Programmes* (New York: Cambridge University Press, 1980), Ernst Nagel, *The Structure of Science: Problems in the Logic of Scientific Explanations* (Indianapolis, In: Hackett Publishing Company, 1979), Trevor J. Pinch, *Confronting Nature: The Sociology of Solar-Neutrino Detection* (Dordrecht: D. Reidel, 1986), Andrew Pickering, *Constructing Quarks; A Sociological History of Particle Physics* (Chicago: The University of Chicago Press, 1984), and Andrew Pickering (ed.), *Science as Practice and Culture* (Chicago: The University of Chicago Press, 1992).

10 See sections 6-10 in, "On the Uses and Disadvantages of History for Life" in Friedrich Nietzsche, *The Untimely Meditations* (Scotts Valley, CA: CreateSpace Independent Publishing Platform, 2016).

11 Ian Hacking claims that certain effects (like the Hall effect) do not exist in nature outside of certain kinds of apparatus. While passing a current through a conductor at right angles to a magnetic field will produce a potential anywhere in nature, nowhere outside the laboratory (the technological devices we construct), is there such a pure arrangement. See Ian Hacking, *Representing and Intervening* (New York: Cambridge University Press, 1987), 225 ff.

12 Thomas Kuhn originally used the metaphor of a map to describe how scientific paradigms shape and define the research programs of scientists. In the case of high-energy physics, the paradigm includes

challenges of doing science-as-practice to produce science-as-knowledge within the institutional context of a megascience laboratory.[13] But to explain how I got involved in the day-to-day realities of doing fixed-target experiments at Fermilab, and how this became the foundation of my research into the social structures of high-energy physics experiments, I need to back up a little, and start at the beginning.

✦ ✦ ✦

I was diagnosed with ADHD at six years old, with parents who refused to medicate me for fear of others' perceptions. I hated school, graduated in the bottom tenth of my high school class, and spent the next four years working as a professional musician, playing clubs in the New York Metropolitan area. The pastor at the church I was attending told me about a college that catered to older students who felt called to the ministry, where grades and SAT scores didn't matter. I applied and was accepted to Nazarene Bible College (N.B.C.) in Colorado Springs. I packed as much as I could in a duffle bag and boarded a flight to Colorado. As I landed on November 28, 1974, I had no idea that seventeen days earlier on Monday, November 11, 1974, a revolution had started in high-energy physics with the discovery of the charm quark, a missing piece of what became the Standard Model of the universe.[14] I had never heard of high-energy physics or quarks. My high school science textbooks taught that the universe was made of atoms that were made of protons, neutrons, and electrons. Six years later, I was

---

the theoretical and experimental evidence that is the basis of the Standard Model, plus the types of experimental methods, detectors, data analysis, and practices described in this book and elsewhere. See Thomas Kuhn, *The Structure of Scientific Revolutions*, 2nd edition (Chicago, IL: The University of Chicago Press, 1970), 109.

13 The physics and detector driven model of the social structure of experiment strings, within the megascience laboratory called Fermilab is also discussed in Lillian Hoddeson, Adrienne Kolb, and Catherine Westfall, *Fermilab: Physics, the Frontier and Megascience* (Chicago: The University of Chicago Press, 2008), 262-280.

14 The J/Psi, charmed, particle was co-discovered at Brookhaven National Laboratory and the Stanford Linear Accelerator Center in 1974. See Min Chen, "The Discovery of the J Particle and the Search of the Charmed Particles" in *Weak and Electromagnetic Interactions at High Energies* (Princeton, NJ: Springer Nature, 1975), 249-295, Wolfgang Panofsky, "Particle Discoveries at SLAC" in *Science*, September 26, 1975 (Vol. 189, 1045-1046), and Michael Riordan, *The Hunting of the Quark: A True Story of Modern Physics* (New York: A Touchstone Book, 1987).

working as an Operations Crew Chief in the Proton Area at Fermilab. But I'm getting ahead of myself.

I graduated from N.B.C. in 1978 with a three-year associate's degree in biblical studies and went directly to MidAmerica Nazarene University where I received a bachelor's degree in religion, then on to graduate school at Wheaton College in 1980 for a master's degree in biblical studies. I finished my first of two years at Wheaton *magna cum laude*, but for reasons I describe elsewhere, I had an existential crisis of faith.[15] I stopped attending graduate school and church and began deconstructing my theology, pulling it apart and reflecting on each element, piece by piece. I embraced my doubts as earnestly seeking pedagogues that would lead me to a new place in my faith and life. I vowed that the slow and arduous reconstruction process would have only one criterion. I would refuse to accept anything that I could not affirm with full integrity.

I needed a job, so I went to Wheaton's student employment office and found a six-month assignment as a "rent-a-tech" supporting Fermilab physicists, engineers, and technicians who worked in the fixed-target experimental beamlines in the Proton Department. I was fascinated by the sheer scale of the science done at the laboratory and wanted to learn about it. Fermilab accelerated protons to energies of 400-800 GeV in the four-mile circular ring of magnets shown on the front cover. Operations Crew Chiefs in Fermilab's main control room extracted about $5x10^{13}$ protons from the accelerator and transported high-energy proton beams to three fixed target experimental areas, about two miles downstream, also shown in the cover. The Pagoda-shaped building pictured at the beginning of this chapter was the Proton Operations control room where Operations Crew Chiefs took custody of the proton beam upstream and used hundreds of dipole and quadrupole magnets to steer and "tune" the assigned beam allocations to primary production targets in the P-West, P-Center, P-East, and Broadband beamlines.

The Proton Crew Chief was responsible for orchestrating the functions of more than a thousand magnets, beamline components, plus enclosure interlock and radiation safety systems using a 16-bit Lockheed MAC-16 minicomputer located in the Pagoda.[16] The MAC-16 was the front-end processor for a

15 See Mark Bodnarczuk, *A Biblical Approach to Jung's Psychology: Experiencing the Self— the Image of God Within* (Research Gate, December 7, 2025), 1-29.

16 See H. R. Barton, *Network Aspects of the Fermilab Control System*, Fermi National Accelerator Laboratory, TM-735 0880.000 March 1977.

CAMAC (Computer Automated Measurement and Control) serial highway that served as the foundation of the Proton Area's beamline control system.[17] The configuration enabled up to 256 CAMAC crates to be placed throughout the Proton Area, and each crate housed a Motorola 6800 microprocessor that functioned as the master crate controller. The crates contained specialized CAMAC modules that read real-time status bits from magnets and beamline components, and enabled a Crew Chief to issue real-time CNAF (Crate, Node, Address, Function) commands that block transferred data to CAMAC serial crate controllers and associated beamline devices from the MAC-16 computer in the Proton Area control room of the Pagoda.

The MAC-16 had 64K of RAM memory, and a limit of 999 devices that could be defined in the database, each of which was connected to a specialized CAMAC card in a CAMAC crate located in service buildings throughout the Proton Area.[18] The Proton Area beamlines were divided up into numbered enclosures and each device in an enclosure was assigned a "name" in the MAC-16 database based on a naming convention that included the device's *location* in an enclosure, the device *type* (e.g., dipole or quadrupole magnets, collimators), and the device's *function* (e.g., as a critical device for radiation safety). The original settings for beamline components and the optics or "tune" for a beamline was established by physicists using two computer programs called TRANSPORT and TURTLE.[19] The values associated with the beam components were subsequently stored as software programs in the MAC-16 database. When it was time to "bring the beam up" after a regular accelerator

---

17 See E. J. Barsotti, *Auxiliary/Master Microprocessor CAMAC Crate Controller*, Fermilab TM-588 0880.000, June 1975.

18 For example, a CAMAC 150 module was used to read the status of, and to control TransRex500 power supplies that were connected to dipole magnets in the beamline enclosure, see R. A. Winje, P. J. Reardon, E. R. Schoop, and W. A. Manahan, *Power Supplies for Magnets*, National Accelerator Laboratory, 1971. CAMAC 064 modules were used to read the status of, and to control the motors that moved large steel collimators in and out of the beamline as part of various beam configurations and radiation safety systems,

19 TRANSPORT was used to establish, design, and tune the beamline optics in the Proton Area, including establishing the required quadrupole and dipole magnet field strengths, and the spacing needed between magnets to guide the proton beam to primary production targets. Once a beam optics design was developed, a Monte Carlo program called TURTLE was used to simulate the performance of the beamline, including beam profiles and beam acceptance. For a description of both software packages, see David C. Carey, *High Energy Charged Particle Optics Computer Programs*, Fermilab-Conf-80/76 2041.000, 1980.

maintenance shutdown, beamline physicists would run the software program and work with the Proton Crew Chief to ensure these values were restored to all beamline devices, with the goal of establishing an optimal tune that consisted of steady beam transport and minimal beam loss.

What I didn't know then that I learned only later is that modern-day high-energy physics experiments call upon a large support structure of craftsmen, technicians, and engineers to design, fabricate, and install the seemingly unrelated devices that compose the primary, secondary, and tertiary beamlines at Fermilab. The Crew Chiefs who actually operate the beamline systems, and the beamline physicists, engineers, and technicians throughout the laboratory are the "unsung" heroes behind every physics publication.[20] The orchestration and integration of Fermilab's support structures and operations personnel, with the experimental physicists (who design, fabricate, install, and operate sophisticated detectors, then collect and analyze the data that are published in journals) are, collectively, what I refer to as science-as-practice, with the final publications of physics results being science-as-knowledge.

Once approved by the Fermilab directorate, experiments ran 24x7. Part of my responsibility when I first started at Fermilab was to work midnight and weekend shifts to assist Crew Chiefs in operating the Proton Area. They modeled the processes involved in monitoring and adjusting the beamline devices used to locate and optimize the transport of primary, secondary, and tertiary beams of various particle types to experimental apparatus located downstream of the Pagoda.[21] They slowly trained me to monitor beam positions on SWICs and to know which magnets were upstream, and which loss monitors were downstream, so when the beam was mistuned I could use the MAC-16 computer to move the beam position in the horizontal or vertical plane and restore an optimal beam tune. I asked questions, took copious notes, studied what I documented off-hours, and I slowly learned the overall configuration of the Proton Area beamline enclosures and the positions and functions of the devices they contained.

20 I describe this more fully in Mark Bodnarczuk, *Research Division Shutdown Highlights: Craftsmen and Scientists at Work*, Fermi National Accelerator Laboratory, Fermilab-Pub-86-183, 1986, 8-9.

21 The primary devices used to locate the particle beam were segmented wire ionization chambers (SWICs), combined with loss monitors that indicated when a beam was mistuned, i.e., scraping magnets and other devices, interacting in the steel, and producing ionized radiation that was detected by loss monitor devices located on or near those beamline components.

My curiosity piqued, I also read entry-level books on the history of high-energy physics and cosmology, which placed the research done at Fermilab in a broader historical and scientific context.[22] Fermilab was studying the very smallest and the very largest objects in the universe. On the small end of the spectrum, I learned that every proton in the universe, including those in my body and all biological life, was composed of two up quarks and a down quark. Every neutron contained two downs and one up quark. The experiments in the Proton Area where I worked, and in Fermilab's Meson and Neutrino Areas, were mapping out the fundamental constituents of the visible universe and the forces by which they interact. This included the 1977 discovery of the Upsilon particle as evidence for the existence of the bottom quark by Leon Lederman's experiment E-288 in the P-Center beamline of the Proton Area—an achievement that earned him the 1988 Nobel Prize in physics.[23]

On the cosmic end of the scale, I learned that Hubble's Law explained how the observed red shift of light from galaxies indicated that the universe was expanding from a central point at speeds greater than the speed of light, and evidence from large telescopes revealed that the universe contained more galaxies than people on the planet. The accidental discovery of Cosmic Microwave Background was an early snapshot of the universe when it was about 380,000 years old and its temperature was three degrees Kelvin.[24] All this, and more, was empirical evidence for the fact that the universe came into existence (*ex nihilo*) about 13.8 billion years ago in a Big Bang.[25] The

---

22 For example, James Trefil, *From Atoms to Quarks: An Introduction to the Strange World of Particle Physics* (New York: Charles Scribner's Sons, 1980), Heinz Pagels, *The Cosmic Code: Quantum Physics as the Language of Nature* (New York: Simon and Schuster, 1982), and Nick Herbert, *Quantum Reality: Beyond the New Physics* (Garden City, New York: Anchor Press, 1985).

23 See Leon Lederman, "The Discovery of the Upsilon, Bottom Quark, and B Mesons" in Lillian Hoddeson, Laurie Brown, Michael Riordan, and Max Dresden (eds), *The Rise of the Standard Model: Particle Physics in the 1960s and 1970s* (New York: Cambridge University Press, 1997), 101-113, and Lillian Hoddeson, Adrienne Kolb, and Catherine Westfall, *Fermilab: Physics, the Frontier, and Megascience* (Chicago: University of Chicago Press, 2008), 191-193.

24 See Rhodri Evans, *The Cosmic Microwave Background: How It Changed Our Understanding of the Universe* (Princeton, NJ: Springer, 2014), and Amedeo Balbi, *The Music of the Big Bang: The Cosmic Microwave Background and the New Cosmology* (Princeton, NJ: Spinger, 2008).

25 See Simon Singh, *Big Bang: The Origin of the Universe* (New York: Harper Perennial, 2005), Steven Weinberg, *The First Three Minutes: A Modern View of the Origin of the Universe* (New York: Basic Books, 1993), and Stephan Hawking, *A Brief History of Time: From the Big Bang to Black Holes* (New York: Bantum, 1988).

Fermilab Theoretical Astrophysics Group pioneered the inner space/outer space connection by applying particle physics knowledge to cosmological problems like the early universe, cosmic microwave background, and dark matter, and the early workshops they conducted helped to establish a new paradigm called particle astrophysics.[26]

After six months, I applied for a job as an Assistant Crew Chief in the Proton Area working for Jon Hawkins, the Operations Group Leader. A panel of physicists and Crew Chiefs who I had worked with interviewed me for the job. One of the physicists, Joel Butler, asked me, "Mark, you have all these degrees in biblical studies. Why do you want a job at Fermilab." I responded without hesitation. "I want a job that if my beliefs change, I don't have to worry about losing my job." Joel and the other interview panel members said nothing, but they understood. The Crew Chiefs I supported on shift had seen my enthusiasm, curiosity, and desire to learn, and I got the job.

Within a year, I was promoted to an Operations Crew Chief and I codified the notes I'd taken into a training manual that I subsequently used to train the Assistant Crew Chiefs and technicians that ran shifts with me.[27] Over the next three years, I documented other important Proton Area systems and activities that Crew Chiefs were responsible for operating. This included the basic principles of high-energy physics beam optics, key issues related to transporting 400 and 800 GeV protons to primary production targets, and an overall guide to the CAMAC serial link and the CAMAC modules used to read the status of beamline devices and to control them.[28]

In 1985, Ken Stanfield moved me to the Research Division Office, and then to the Fermilab Director's Office as an institutional program manager. I decided to go back to graduate school at the University of Chicago as part of the *Committee on the Conceptual and Historical Studies of Science*, where I studied the conceptual development of physics from Babylonian astronomy to quantum mechanics. I was fascinated by the fact that the

26 See Edward W. Kolb, Michael Turner, David Lindley, Keith Olive, and David Seckel (eds), *Inner Space/Outer Space: The Interface Between Cosmology and Particle Physics* (Chicago: University of Chicago Press, 1986).

27 See Mark Bodnarczuk, *The Experimental Areas Operations Training Manual for New Employees*, Fermi National Accelerator Laboratory, September 1982.

28 See Mark Bodnarczuk, *The Experimental Areas Beamline Transport System*, Fermi National Accelerator Laboratory, June, 1983, and Mark Bodnarczuk, *An Operational Approach to High-Energy Physics Detectors at Fermilab*, Fermi National Accelerator Laboratory, TM-1368 2000.000, 1985.

search for the fundamental constituents of the universe and the forces by which they interact is as old as Western civilization.[29] Aristotle (384–322 BC) considered Democritus (460-370 BC) to be the father of atomism.[30] Democritus argued that if you continue dividing matter into smaller and smaller parts, you will eventually reach a fundamental, indivisible unit called an *atomos* (ἄτομος), meaning that which is indivisible. He also claimed that atoms were eternal, unchangeable, indestructible particles of differing shapes (smooth, hooked) that are in constant motion in an infinite void. The visible changes in the physical world that we observe are caused by atoms colliding, rebounding, or rearranging, rather than changes to the atoms themselves.[31]

An underlying question that natural philosophers and scientists sought an answer to over the last 2,500 years is whether atoms are "real" objects, or whether they are philosophical or mathematical constructs that have no objective existence in the world. It's hard to believe that we were still asking that question about atoms at the beginning of the twentieth century. In 1900, the empirical existence of atoms was a subject of heated scientific debate, with scientists like Ernst Mach and Wilhelm Ostwald denying their existence.[32] Einstein's 1905 theoretical work on Brownian motion created a mathematical framework for verifying the objective existence of atoms.[33] He derived a formula that related the mean square displacement of particles to temperature, viscosity, and particle size, enabling him to calculate

29 See G. E. R. Lloyd, *Early Greek Science: Thales to Aristotle* (New York: W. W. Norton, 1974), G. E. R. Lloyd, *Greek Science After Aristotle* (New York: W. W. Norton, 1975), and David Lindberg, *The Beginnings of Western Science: The European Scientific Tradition in Philosophical, Religious, and Institutional Context, Prehistory to A.D. 1450* (Chicago, IL: University of Chicago Press, 2008).

30 See Aristotle, *On Generation and Corruption Book I*, and *Metaphysics* in Richard McKeon (ed), *The Basic Works of Aristotle* (New York: Random House, 1941), 470 ff. and 689 ff.

31 See David Furley, *The Greek Cosmologists: volume 1, The Formation of the Atomic Theory and its Earliest Critics* (New York: Cambridge University Press, 2006), and David Furley, *Two Studies in the Greek Atomists* (Princeton, NJ: Princeton University Press, 1967).

32 See J. T. Blackmore (ed), *Ernst Mach: A Deeper Look: Documents and New Perspectives* (New York: Springer, 1992), and David Lindley, *Boltzmann's Atom: The Great Debate That Launched a Revolution In Physics* (New York: Free Press, 2016).

33 See Albert Einstein, *Investigations on the Theory of the Brownian Movement* (Garden City, New York: Dover Publications, 1956), Luis Navarro Veguillas, *The Lesser-Known Albert Einstein: Without a Trace of Relativity* (Princeton, NJ: Springer, 2023), and John Stachel (ed), *Einstein's Miraculous Year: Five Papers That Changed the Face of Physics* (Princeton, NJ: Princeton University Press, 1998).

Avogadro's number, and by 1908 he suggested a way to test his theory experimentally. Later that year, Jean Perrin conducted a series of experiments that confirmed Einstein's predictions and finally established the objective reality of atoms.[34]

Until 1967, high-energy physicists were asking that same question about the objective reality of quarks.[35] Proposed in 1964, by Murray Gell-Mann and George Zweig, to explain the internal structure of hadrons,[36] and predicted in 1966 by James Bjorken's hypothesis of "scaling,"[37] the 1967 point-like structure in hadrons experiment conducted at SLAC by Jerome Friedman, Henry Kendall, and Richard Taylor was the first empirical evidence that quarks were real objective entities in the world. By firing 8-20 GeV electrons at hydrogen and deuterium targets they detected electrons scattering at large angles,

34 See Jean Perrin, *Brownian Movement and Molecular Reality* (New York: Legare Street Press, 2022), and George Smith and Raghav, *Brownian Motion and Molecular Reality* (New York: Oxford University Press, 2020).

35 For a relatively recent dialogue about the objective reality of quarks published after their discovery by high-energy physicists, see Andrew Pickering, *Constructing Quarks; A Sociological History of Particle Physics* (Chicago: The University of Chicago Press, 1984) for the social constructionist view. For the objective realist view, philosopher of science, Ian Hacking, discusses the 1977 Stanford University experiment performed by LaRue, Fairbank, and Hebard that measured fractional charges on a cryogenically cooled niobium ball sprayed with positrons. Hacking concludes his analysis by saying, "From that day forth I've been a scientific realist. *So far as I'm concerned, if you can spray them, then they are real.*" See Ian Hacking, *Representing and Intervening* (New York: Cambridge University Press, 1987), 22-23. Also see, George S. LaRue, William M. Fairbank, and Arthur F. Hebard, "Evidence for the Existence of Fractional Charge on Matter," in *Physical Review Letters* (Vol. 38, Issue 18), May 2, 1977.

36 Initially, Gell-Mann and Zweig's model contained three quarks, i.e., up, down, and strange, although they speculated about the existence of a fourth quark. See George Zweig, *Discovering Quarks: Remembering Feynman, Gell-Mann, and Tollestrup* (New York: Cambridge University Press, 2025), and Harold Fritzsch and Murray Gell-Mann (eds), *50 Years of Quarks* (Hackensack, NJ: World Scientific Publishing Company, 2015).

37 Bjorken's hypothesis of "scaling" in deep inelastic scattering predicted that high energy electrons would scatter off nearly free-moving point-like constituents in hadrons, paving the way for Richard Feynman's parton model in 1969, and subsequently the quark-gluon model of Quantum Chromodynamics (QCD) on the 1972-1973 timeframe. See James D. Bjorken, "Asymptotic Sum Rules at Infinite Momentum" in *Physical Review*, 1969, 179, 1547, James D. Bjorken "Deep-Inelastic Scattering: From Current Algebra to Partons" in Lillian Hoddeson, Laurie Brown, Michael Riordan, and Max Dresden (eds), *The Rise of the Standard Model: Particle Physics in the 1960s and 1970s* (New York: Cambridge University Press, 1997), 589-599, and Michael Peskin, *Concepts of Elementary Particle Physics* (Cambridge: Cambridge University Press, 2019). Feynman describes his parton model in Richard Feynman, *Photon-Hadron Interactions* (Boca Raton, FL: CRC Press, 1998), which is also described in Michael Riordan, *The Hunting of the Quark* (New York: Simon and Schuster, 1987), 148-155.

indicating they were hitting smaller, hard, charged constituents inside of protons and neutrons.[38]

The existence of a fourth quark (charm) was predicted by Sheldon Glashow and James Bjorken in 1964 as a way to restore symmetry between the known leptons and quarks at the time, establishing a pattern of two weak doublets for each.[39] As mentioned previously, this was followed by the 1974 discovery of the charm quark at SLAC and Brookhaven, which helped define the research program for a number of Fermilab experiments, including the experiment string that used the Tagged Photon Magnetic Spectrometer described in this book.[40]

Given my background as an Operations Crew Chief, my role as Fermilab's first Institutional Quality Assurance Program Manager, and my graduate work at the University of Chicago, I became interested in the nature of scientific practice and the historical development of "big science" into the type of "megascience" described in chapter 2 of this book, and in Lillian Hoddeson's book, *Fermilab: Physics, the Frontier and Megascience*.[41] I was fascinated by

---

38 See Michael Riordan, *The Hunting of the Quark* (New York: Simon and Schuster, 1987), 130-135, Jerome Friedman "Deep-Inelastic Scattering and the Discovery of Quarks" in Lillian Hoddeson, Laurie Brown, Michael Riordan, and Max Dresden (eds), *The Rise of the Standard Model: Particle Physics in the 1960s and 1970s* (New York: Cambridge University Press, 1997), 566-588, and Henry W. Kendall "Deep Inelastic Scattering: Experiments on the Proton and the Observation of Scaling" in *Rev. Mod. Phys.* 63, 597 (1991).

39 In 1964, the known fundamental particles included four leptons (electron, electron neutrino, muon, and muon neutrino) and three proposed quark flavors (up, down, and strange) that explained the internal structure of hadrons. See Sheldon Glashow, *The Charm of Physics* (New York: Touchstone Books, 1991).

40 Experiment E-516, the Tagged Photon Magnetic Spectrometer at Fermilab, was formally proposed in June 1976 when the team submitted their "Letter of Intent" to the directorate. The official proposal for the experiment (P-516) was finalized and addressed by October 3, 1977, with approval for E-516 granted on November 15, 1977.

41 Historians do not agree on when big science began. Historians of technology often point to the big science done circa 1900 in early industrial laboratories, such as General Electric or Du Pont, e.g., George Wise, *Willis R. Whitney, General Electric and the Origins of U.S. Industrial Research* (New York: Columbia University Press, 1985) or David A. Hounshell and John Kenly Smith, Jr., *Science and Corporate Strategy: Du Pont R&D, 1902–1980* (Cambridge: Cambridge University Press, 1988). In particle physics it is accepted that big science emerged in nuclear physics research at Berkeley during the 1930s at the time of Ernest Lawrence's pioneering cyclotrons: J. L. Heilbron and Robert Seidel, *Lawrence and His Laboratory: A History of the Lawrence Berkeley Laboratory*, vol. 1 (Berkeley: University of California Press, 1989). Weapons historians look later at laboratories organized during or just after the Second World War: L. Hoddeson, P. Henriksen, R. Meade, and C. Westfall, *Critical*

the relentless search for new physics at higher and higher energies, requiring detectors with larger fiducial volumes, increasingly complex configurations of detector subsystems with tens of thousands of channels of electronics, and more powerful online and offline computing power for recording and reconstructing particle interactions that occurred in detectors like the Tagged Photon Magnetic Spectrometer.

The thousands of hours that I spent on shift as a Crew Chief working hand-in-hand in partnership with hundreds of physicists and graduate students caused me to reflect on the social structure and social interactions in these collaborations. My time in the Fermilab Director's Office provided insights into the institutional resource economies that physicists had to trade in to perform their experiments within the organizational complexities of laboratory life. My role as the Fermilab correspondent to the *CERN Courier* and as a contributor to the *Fermilab Report* became a springboard to publishing articles about the overall experimental programs in the Proton, Meson, and Neutrino areas.[42] I met Lillian Hoddeson, the Fermilab Historian, on the 1986 timeframe. She taught me the value of preserving the history of high-energy physics and invited me to join the Fermilab History Collaboration.[43]

In chapters 3 and 4 of this book, I discuss how the physics and detector driven model of experiment strings applied to Fermilab's CDF and DO collider detectors that were operational at the time. I also tried to look ahead to how the physics and detector driven model might apply to, what, at that time, was the next generation of collider detector collaborations, e.g., the Solenoidal Detector Collaboration (SDC) at the proposed Superconducting Super Collider (SSC) Laboratory.[44] As the physics and

Assembly: A History of Los Alamos during the Oppenheimer Years, 1943–1945 (New York: Cambridge University Press, 1993).

42 See Mark Bodnarczuk, "Decommissioning of the 15-Ft Bubble Chamber Marks the End of an Era" in *Fermilab Report*, March/April 1988, 15-22, and Mark Bodnarczuk, "Workshop on Physics at Fermilab in the 1990's" in *Fermilab Report*, October 1989, 12-15.

43 My work with the Fermilab History Collaboration and the physics and detector driven model of the social structure of experiment strings is discussed in Lillian Hoddeson, Adrienne Kolb, and Catherine Westfall, *Fermilab: Physics, the Frontier and Megascience* (Chicago: The University of Chicago Press, 2008), 262-280.

44 The SSC Laboratory was officially cancelled following a House of Representatives vote to cut the funding on October 19, 1993, with President Bill Clinton signing the bill terminating the project on October 30, 1993, after about 20 percent of the project was finished and $2 billion dollars had been spent. See Michael Riordan, Lillian Hoddeson, and Adrienne Kolb, *Tunnel Visions: The Rise and Fall of*

detector driven model predicted, the higher energies and complexities of the proposed physics measurements at 40 TeV center-of-mass proton-on-proton collisions demanded that the SDC's design would be among the largest, most costly, and complex devices ever designed for any scientific activity. The 1990 estimates by the SDC collaboration described a detector that would cost $500 million (in 1990 dollars), weigh 10,000 tons, and have 20 million channels of electronics.[45] In terms of the "people design" of the SDC, there were about 600 physicists in the collaboration, with a higher-level social structure consisting of one Spokesperson, three Deputy Spokespersons, a Technical Manager, and a 13 member Executive Board.[46] The collaboration had a written and ratified Governance Document which described the organization's structure and the management of the group, as well as the rules for admitting additional members into the experiment.[47]

Looking back, it is important to remember that the articles included in this book were published over 30 years ago, and although predicted by the Standard Model at that time, the top quark and Higgs Boson had not been discovered.[48] The science-as-knowledge and science-as-practice of high-energy physics has

the Superconducting Super Collider (Chicago, IL: The University of Chicago Press, 2015).

45 For details, see the *Solenoidal Detector Collaboration Expression of Interest*, 24 May 1990.

46 The higher-level social structure of CERN's Delphi experiment at LEP, was referred to as a "parliament." The collaboration consisted of over 500 physicists and was headed by a Collaboration Board which is responsible to reconcile potential conflicts of interest between participating institutions and overview the final production of physics results. In addition, a collaboration Executive Committee was responsible to make technical recommendations for the experiment which had to be subsequently ratified by the Collaboration Board. See Gordon Fraser, "Delphi," in *CERN Courier*, vol. 30, no. 8, November 1990, 1-5.

47 For details, see the *Solenoidal Detector Collaboration Expression of Interest*, 24 May 1990, 93 ff.

48 The discovery of the top quark by the CDF and D0 collaborations was announced on March 2, 1995 with a mass of about 173 GeV. For a discussion of the discovery of the top quark, see Paul Tipton and Tony Liss, "The Discovery of the Top Quark" in *Scientific American Magazine* Vol. 277 No. 3 (September 1997), p. 54, Claudio Campagnari and Melissa Franklin, "The Discovery of the Top Quark" in *Rev. Mod. Phys.* 69, 137, January 1997, and Lillian Hoddeson, Adrienne Kolb, and Catherine Westfall, *Fermilab: Physics, the Frontier, and Megascience* (Chicago: University of Chicago Press, 2008), 348-349. The discovery of the Higgs Boson at CERN by the ATLAS and CMS collaborations was announced on July 4, 2012, with a mass of about 125 GeV. For a discussion on the discovery of the Higgs Boson, see Jon Butterworth, *Most Wanted Particle: The Inside Story of the Hunt for the Higgs, the Heart of the Future of Physics* (New York: The Experiment, 2016), Sean Carroll, *The Particle at the End of the Universe: How the Hunt for the Higgs Boson Leads Us to the Edge of a New World* (New York: Dutton, 2013), and Mario Campanelli, *Inside CERN's Large Hadron Collider: From the Proton to the Higgs Boson* (Hackensack, NJ: World Scientific, 2015).

come a long way since then. Today, the Standard Model consists of seventeen fundamental particles and three of the four fundamental forces.[49] What was perceived in 1990 as an unprecedented level of scale, cost, and complexity with the proposed SDC detector is now dwarfed by the ATLAS and CMS detectors currently running at the Large Hadron Collider (LHC) at CERN.[50] Performing the type of historical, philosophical, and sociological research done on the E-516-691-769-791 experiment string in this book for the ATLAS or CMS experiments would be a daunting task, requiring a group of historians, philosophers of science, and sociologists working for an extended period of

---

49 This includes: a) twelve Fermions that compose matter divided into six quarks (up, down, charm, strange, top, bottom) and six leptons (electron, electron neutrino, muon, muon neutrino, tau, tau neutrino), b) four Bosons which are the four particles that mediate the interactions of the electromagnetic force (photons), the strong force in hadrons (gluons), and the weak force (W and Z bosons), and c) the Higgs Boson which is the particle that connects to the Higgs Field and gives mass to all the other fundamental particles. Finalized in the mid-1970s and confirmed by the 1995 discovery of the top quark, and the 2012 discovery of the Higgs Boson, the Standard Model is a "periodic table" for subatomic particles. The fourth force, gravity, is not currently part of the Standard Model, but a massless particle called the graviton is the proposed carrier of the gravitational force. If detected experimentally, the graviton could reconcile the quantum field theory of the Standard Model (i.e., Quantum Chromodynamics, Electroweak Theory, and the Higgs Mechanism combined into a Lagrangian mathematical framework), with Einstein Field Equations (EFE) that relate the geometry of spacetime to the distribution of mass and energy within it. The Einstein Field Equations model gravity, not as a force, but rather as a four-dimensional curved, continuous *spacetime fabric* produced by the density of mass and energy as described by Einstein's 1915 theory of General Relativity. See Robert Geroch, *General Relativity from A to B* (Chicago: The University of Chicago Press, 1981), and Leonard Susskind and Andre Cabannes, *General Relativity: The Theoretical Minimum* (New York: Basic Books, 2023).

50 The ATLAS detector has over 100 million channels of electronics that are distributed across the Pixel Detector, Semiconductor Tracker (SCT), and the Transition Radiation Tracker (TRT). The ATLAS collaboration consists of over 6,000 physicists, engineers, and PhD graduate students from over 100 countries and more than 180 institutions, with about 3,000 scientific authors. In terms of the people-design, the ATLAS Collaboration has a massive, democratic, decentralized organization led by: a) an elected Spokesperson, b) a Management Team that oversees all aspects of the experiment, including data taking, detector operation, and analysis, c) a Collaboration Board that represents the participating institutes, elects the Spokesperson and makes high-level strategic decisions, d) working groups that analyze data and develop detector technologies, and e) Project Offices and Coordinators with specialized roles for managing long-term improvements and upgrades to the detector. See Dan Green (ed), *At the Leading Edge: The ATLAS and CMS LHC Experiments* (Hackensack, NJ: World Scientific Publishing, 2010), The Atlas Collaboration, *ATLAS: A 25-Year Insider Story of the LHC Experiment* (Hackensack, NJ: World Scientific Publishing, 2019), and Anton Radevsky and Emma Sanders, *Voyage to the Heart of Matter: The Atlas Experiment at CERN* (Winterbourne, Berkshire, UK: Andreas Papadakis Pub, 2010).

time.[51] But I believe that using a physics and detector driven analysis of the social structure of ATLAS and CMS would be equal to the task, providing an empirical foundation against which historical, philosophical, and sociological research on the next generation of accelerators, detectors and collaborations at the proposed Future Circular Collider (FCC) at CERN could be benchmarked.

If approved in 2028, the FCC would be a 57-mile ring at an average depth of 656 feet underground that crosses under Lake Geneva and straddles the French Swiss border. It would have eight surface sites, four underground experimental halls, and would cost about $20 billion. The FCC would run in two stages. Stage 1 would be an electron-positron lepton collider designed as a Higgs, electroweak, and top quark factory, operating at center-of-mass energies ranging from 90 GeV to 365 GeV, with operation beginning between 2045 and 2050. Stage 2 would be a proton-proton hadron collider with a center-of-mass energy of 100 TeV, about 30 times higher than the 14 TeV center-of-mass energy of the LHC. It will be designed to study dark matter and to search for new particles with masses up to several tens of TeV. Stage 2 would utilize the same tunnel as Stage 1 and begin operating in the 2070s.[52]

Three of the four proposed FCC detectors are described in the recently issued feasibility study (the IDEA, CLD, and ALLEGRO detectors), each of which would be unprecedented in scale, cost, and complexity.[53] For comparison, at their peak, the CDF and D0 collaborations each had about 600

---

51 For some attempts to-date to study the ATLAS and CMS detectors and collaborations see Dominique Raynaud, "Collaborative Networks in Particle Physics: A Sociological Inquiry into the ATLAS and CMS Collaborations" in *Sociology of Science and Technology* 7:169-193 (2016), Peter Jenni, Tejinder S. Virdee, Ludovico Pontecorvo, and Shantha Liyanage "Chasing Success: The ATLAS and CMS Collaborations" in Shantha Livanage, Markus Nordberg, and Marilena Streit-Bianchi (eds), *Big Science, Innovation, and Societal Contributions: The Organisations and Collaborations in Big Science Experiments* (Cambridge: Oxford University Press, 2024), 22-55, and Zeeya Merali, "The Large Human Collider: Social Scientists Have Embedded Themselves at CERN to Study the World's Biggest Research Collaboration" in *Nature*, vol. 464, no. 7288, 25 Mar. 2010, 482 ff.

52 See "Future Circular Collider Feasibility Study Report, Volume 1, Physics, Experiments, Detectors" in *The European Physical Journal C*, Vol. 85, Article 1468, (2025), "Future Circular Collider Feasibility Study Report, Volume 2, Accelerators, Technical Infrastructure and Safety" in *The European Physical Journal ST*, Vol. 234, 5713-6197, (2025), and "Future Circular Collider Feasibility Study Report, Volume 3, Civil Engineering, Implementation, and Sustainability" in *The European Physical Journal ST*, Vol. 234, 5113-5383, (2025).

53 See "Future Circular Collider Feasibility Study Report, Volume 1, Physics, Experiments, Detectors" in *The European Physical Journal C*, Vol. 85, Article 1468, (2025), 78-116 and 126-154.

physicists with detectors consisting of about one million channels of electronics.[54] The ATLAS collaboration includes about 6,000 physicists, engineers, technicians, and support staff with the detector having about 100 million channels of electronics.[55] One could imagine that each of the FCC collaborations might have as many as 12–15,000 physicists, engineers, technicians, and support staff with detectors consisting of one billion channels of electronics.

What is clear from the FCC feasibility study is that the push to higher and higher energies will continue to drive the design of high-energy physics detectors, and this in turn will shape and define the social structures of these collaborations in ways that echo this Fermilab study. It's important to note that the five resource economies that I describe in the chapters below (i.e., experimental real estate, proton economics, computing economics, physics economics, and physicist economics) are key areas covered in the FCC feasibility study. Given the current size of the high-energy physics community, one of the most problematic economies for FCC experiments may be physicist economics.[56] Taken together, these are indications of the heuristic and staying power of a physics and detector driven approach to studying what will soon become the largest scientific enterprises on the planet.

Mark Bodnarczuk
Breckenridge Institute®
Longmont, Colorado
June, 2026

54 All 600 members of the CDF collaboration were high-energy physicists, which included about 100 postdoctoral researchers and 100 graduate students.

55 About 64 percent of the ATLAS collaboration are high-energy physicists with a significant portion of the remaining membership composed of engineers and technicians who provide essential infrastructure for the detector's maintenance and upgrades, and a small percent of administrative and support staff who manage the collaboration's global operations.

56 Based on data from major international collaborations at CERN, Fermilab, KEK, DESY, the current estimate of the number of high-energy physicists in the world is between 10,000 and 20,000.

Wilson Hall, the fifteen-story central laboratory building at Fermilab, was designed by its first director, Robert Rathbun Wilson, and built between 1971 and 1974. Inspired by the Cathedral of Chartres in France, Wilson's vision was to create a "cathedral of science" with an open atrium design intended to foster scientific collaboration.

1

Some Sociological Consequences of High-Energy Physicists' Development of The Standard Model[57]

In a scientific discipline that went from experiments with less electronics than a videocassette recorder to 10^5 channels, and from collaborations with 5-10 members to 300 during the years 1964-1979, the notion of what high-energy physics is, or what constitutes *being* a high-energy physicist, cannot be viewed simply as an immutable category that is "out there"—that remains fixed despite these and other developments. What high energy physics is as a discipline and what it means to be a high-energy physicist are renegotiated by participants relative to the experimental and theoretical practices of the field at any given time. In this article I will explore some of the sociological consequences of the decisions made by high-energy physicists as they constructed the edifice that has come to be known as the Standard Model.[58]

57 This article was originally published in Lillian Hoddeson, Laurie Brown, Michael Riordan, and Max Dresden (eds), *The Rise of the Standard Model: Particle Physics in the 1960s and 1970s* (New York: Cambridge University Press, 1997), 384-393.

58 Currently, there are numerous approaches to the social study of science. For a traditional view of the sociology of science, see Robert Merton, *The Sociology of Science, Theoretical and Empirical Investigations* (Chicago: The University of Chicago Press, 1973). Some of the earliest work in the sociology of knowledge can be found in Karl Mannheim, *Ideology and Utopia; An Introduction to the Sociology of Knowledge* (New York: Harcourt Brace Jovanovich, 1985), and the early development of the "strong programme" of the sociology of scientific knowledge (SSK) is best represented in David Bloor, *Knowledge and Social Imagery*, 2nd ed. (Chicago; The University of Chicago Press, 1991). Some of the more moderate proponents of SSK include Bruno Latour, *Science in Action: How to Follow Scientists and Engineers through Society* (Cambridge: Harvard University Press, 1987) and Trevor J. Pinch, *Confronting Nature: The Sociology of Solar-Neutrino Detection* (Dordrecht: D. Reidel, 1986). Perhaps the most radical SSK position is in Steve Woolgar, *Science, the Very Idea* (New York: Tavistock

Many of these physicists' decisions about the Standard Model have already been carefully documented in Andrew Pickering's sociological history of the development of particle physics, as well as numerous chapters from this volume.[59] I am thinking particularly of factors like the notion of quarks and the development of the Eightfold Way and S-matrix bootstrap theory; scaling, hard scattering, and the 1967 MIT-SLAC experiment's evidence for point-like structure in hadrons; the quark-parton model that was supported by experimental evidence for J/ψ, bare charm, and upsilon particles; the development of gauge theory, the unified theory of electroweak interactions, with the experimental evidence of neutral currents; and finally the development of a theory of strong interactions—quantum chromodynamics.

Other than to underscore physicists' decisions to pursue higher and higher energies (as evidenced in the construction of a 200, then 400, GeV proton accelerator at Fermilab), I will not recount these details here. Rather, within the context of such decisions, I attempt to describe how the increases in scale, cost, and complexity mentioned earlier were *consequences of the choices* to go to higher and higher energies in response to the experimental evidence and theoretical constructs that emerged from 1964 to 1979.[60] More specifically,

Publications, 1988). More recently, Andrew Pickering has collected a number of essays that focus on the central role of practice in SSK, in Andrew Pickering (ed.), *Science as Practice and Culture* (Chicago: The University of Chicago Press, 1992), and Stephen Cole has provided the first serious critique, by a traditional sociologist, of the SSK position, in Stephen Cole, *Making Science, Between Nature and Society* (Cambridge: Harvard University Press, 1992).

59 Andrew Pickering, *Constructing Quarks; A Sociological History of Particle Physics* (Chicago: The University of Chicago Press, 1984).

60 Pickering claims the relationship between experimental and theoretical research traditions is symbiotic in that each generation of practice within one tradition provides a context within which the succeeding generation of practice in the other finds its justification and subject matter. Peter Galison claims that the truism that "experiment is inextricable from theory" or that "experiment and theory are symbiotic" is useless because, while vague allusions to Gestalt psychology may have been an effective tactic against dogmatic positivism, experimentalists' real concern is not with global changes of worldview. For Galison, the salient issue is where theory exerts its influence in the experimental process and how experimentalists use theory as part of their craft. My point is that once physicists decide to study certain physical phenomena and theoretical constructs at higher and higher energies, such a decision has physical consequences (e.g., larger accelerators given the technologies during the 1964 to 1979 era, and larger more heavily instrumented fiducial volumes in apparatus to detect myriad particle interactions) and sociological consequences of the types that constitute the remainder of this chapter. See Pickering, *Constructing Quarks*, 10-11, Peter Galison, How *Experiments End* (Chicago: The University of Chicago Press, 1987), 245, and Chapter 18, of this volume, see Peter Galison, "Pure and Hybrid Detectors: Mark I and the Psi" in Lillian Hoddeson, Laurie Brown, Michael Riordan, and

one consequence witnessed at Fermilab was the development of an increasingly complex and bureaucratic organizational infrastructure, described below, that included a number of interrelated resource economies, each having its own commodity.[61] Another consequence of larger more complex detectors was the formation of larger, more complicated social structures for the collaborations that designed, fabricated, installed, and operated the experiments, and an increased scale and sophistication of the computing power needed to collect data samples and bring the results to final publication.

After 1972, Fermilab operated the highest-energy particle accelerator in the world, and consequently competition for use of the wide variety of particle beams it produced was intense. In order to gain access to a beamline, experimentalists had to navigate a number of interrelated resource economies that were embedded within an institutional structure headed by a single scientist, the director, who had ultimate authority in all matters scientific and otherwise.[62] Experimentalists had to learn to trade with and for these commodities to participate in the production of knowledge in high-energy physics. Physicists negotiated with these commodities and often fought over them.[63]

One economy at Fermilab was proton economics, with protons as the commodity. The overall magnitude of the economy was limited by such factors as accelerator flux, efficiencies in primary beam transport, cross sections for secondary beam production, secondary beam transport efficiencies, and expected reaction rates in experimental targets. For example, given the cross

Max Dresden (eds), *The Rise of the Standard Model: Particle Physics in the 1960s and 1970s* (Cambridge: Cambridge University Press, 1997), 308-337.

61 I developed the resource economy model from a case study of several Fermilab experiments. See Mark Bodnarczuk, "The Social Structure of Experimental Strings at Fermilab: A Physics and Detector Driven Model," Report No. Fermilab-PUB-91/63, Batavia, March 1990, 2-7. This model is not unlike Bruno Latour's more generic model of cycles of credit that involve conversions of different types of capital (recognition, grant money, equipment, data, arguments, articles, etc.) into the "credibility" that scientists need to make moves within a scientific field. See Bruno Latour and Steve Woolgar, *Laboratory Life: The Construction of Scientific Facts* (Princeton: Princeton University Press, 1986), 187-233.

62 Maurice Goldhaber, in an unpublished article entitled "The Beginning of Program Committees," remarks on the early formation of program committees appointed by laboratory directors for the purpose of obtaining independent assessments of the laboratory's research program.

63 Using numerous case studies, David Hull claims that not only are infighting, mutual exploitation, and even personal vendettas typical behavior for many scientists, but that this sort of behavior actually facilitates scientific development. David Hull, *Science as a Process: An Evolutionary Account of the Social and Conceptual Development of Science* (Chicago: The University of Chicago Press, 1988), 26.

section for neutrino scattering (10^{-36} cm^2) and the pion cross section (10^{-26} cm^2), the decision to approve experiments using incident beams of neutrinos was already a major decision that affected proton economics.[64] A neutrino beam was more costly than a pion beam in terms of the number of protons needed to produce it; this was further complicated if the cross section for event production in the proposed experimental target was low and the experiment required a large number of particle events to be competitive with previously accumulated world samples. Given this intense competition, it became difficult to navigate contentious beam management issues, especially in the kind of user-based environment that typified Robert Wilson's philosophy at Fermilab.[65]

Experimental real estate was a second economy; here the commodity was possession of an experimental hall at the end of a beam spigot to house the collaboration's apparatus. As detectors became larger and more complicated, the lead time needed to assemble and operate apparatus also increased. Consequently, physicists who were granted a piece of experimental real estate and some beam time tended to move into an experimental hall with the explicit goal of performing that experiment, and the implicit goal of not moving out. Gaining access to an experimental hall, especially when the incumbent collaboration was desperately trying to hold its place in line, made gaining possession of this commodity a top priority in Fermilab's user-based environment.

Another economy, which I call "physicist economics," is based on the commodity of physics expertise. The scale and complexity of the experiments during the 1964 to 1979 period continued to increase at an unprecedented rate, but the number of high-energy physicists who could commit to perform experiments was constrained by the total number available at that time and the rate at which new PhD graduate students were being produced. So the enormous increase in the scale and complexity of experiments made physics

64 The pion cross section is roughly constant for energies above 2 GeV at about 40 millibarns. The neutrino cross section is not constant, but is linearly proportional to the energy. For Fermilab, a reasonable neutrino energy to use to calculate this was 100 GeV, which would give a neutrino cross section of about 0.7 picobarns.

65 In Chapter 20 of this volume, Catherine Westfall notes that Edwin Goldwasser (who later became Wilson's Deputy Director) was one of the first to address user discontent in the United States. See Catherine Westfall, "Panel Session: Science Policy and the Social Structure of Big Laboratories" in Lillian Hoddeson, Laurie Brown, Michael Riordan, and Max Dresden (eds), *The Rise of the Standard Model: Particle Physics in the 1960s and 1970s* (Cambridge: Cambridge University Press, 1997), 364-383.

expertise an increasingly valuable commodity.[66] Larger, more complex, and increasingly modularized detectors required larger, more complex, and increasingly modularized social structures with the appropriate number of physicists and the distribution of expertise needed to design, fabricate, install, and operate the apparatus and to develop the computing systems and software programs used to reconstruct and analyze the particle events that were recorded. By the late 1970s, collaborations were characterized by an unprecedented division of labor, so no single member of a collaboration had a detailed knowledge of the entire detector. As pointed out by Galison, this kind of modularization provided each institution with a visible manifestation of its contribution to the experiment.[67] Not only was the modularization of detectors an important aspect of carving out a piece of physics to work on, it was also an important political issue back at the home university. Proposals were increasingly judged on the "physicist design" of the experiment and how well it mapped to the experimental design, with laboratory directors and their advisory committees focusing more and more on whether the collaboration had enough physicist power to make good on its experimental claims.

But the consequences of physicists' choices (increased scale and complexity of detectors, accelerators, and the associated social structures) are most easily seen in a fourth resource economy, computing economics, where online and offline computing power was the sought-after commodity. One example was the attempt to do high-statistics charm production experiments at Fermilab in the late 1970s.[68] On the one hand, the advantages of online data reduction using sophisticated trigger processors had to be balanced against the risk of coming up empty-handed due to incorrect trigger assumptions and the problem of obtaining adequate offline computing power. On the other hand, the more secure approach to online data acquisition (the write-it-all-to-tape approach) had to be balanced against the problem of obtaining immense

66 For example, a recent study of the research program for the 1990s performed by the High Energy Physics Advisory Panel included a detailed demographic study of "manpower considerations" during the time period under study. See the HEPAP Subpanel, "The U.S. High Energy Physics Research Program for the 1990s," Report No. DOE/ER-0453P, Washington, D.C., April 1990, 68ff.

67 Peter Galison referred to the visibility that modularization provided participants in his talk at this Symposium (unpublished).

68 See Bodnarczuk, "The Social Structure of Experimental Strings," for a detailed case study of Fermilab experiments E-516, E-691, E-769, and E-791 that performed high-statistics photoproduction and hadroproduction of charmed particles.

offline computing resources, which was difficult given Wilson's belief that the bulk of computing for experiments should be provided by the collaboration's home institutions.[69] There was an abrupt explosion in the number of channels of electronics required in detectors after 1980. So the period during the development of the Standard Model was the calm before the storm—before the explosion in scale, cost, and complexity of hadron collider detectors (such as CDF) that were conceived after 1977.[70]

A final resource economy was physics economics; the commodity of published physics results was traded back to the laboratory director as a return on investment and was the key to obtaining additional resources to perform follow-up experiments. Within physics economics, the laboratory director's ability to approve or disapprove an experiment was a powerful management tool for leveraging wayward experimenters who failed to make good on their promises, especially when they wanted to move on to the greener pastures of follow-up experiments without first publishing the results of their existing experiment.

The study of various Fermilab experiments mentioned earlier also shows that within this socioeconomic-scientific infrastructure of the laboratory, experiments were performed in a series of follow-up experiments in which an experiment was performed, then transformed into a second, then a third, or a fourth experiment. I call these series of experiments "experiment strings."[71] The key to describing these transformations is the ability to characterize the continuities between individual experiments in a given string. Evidence that emerged from the previously mentioned study suggests that experiment strings exhibit well-defined continuities in the physics goals, detector configuration, and the core group of collaborators that participated in nine or more experiments over

---

69 Mark Bodnarczuk interview with Robert Wilson, 24 September 1992.

70 The UA1 detector at CERN had about 50,000 channels, the CDF and D0 detectors at Fermilab, and the SLD detector at SLAC each had over 100,000 channels, the ALEPH detector at LEP had about 700,000 channels, and the proposed SDC and GEM detectors at the Super Conducting Super Collider (SSC) could have had 50,000,000 channels, depending on the available technology.

71 See Bodnarczuk, "The Social Structure of Experimental Strings," 14-20; Joel Genuth, "Historical Analysis of Selected Experiments at US Sites," *AIP Study of Multi-Institutional Collaborations: Phase I. High Energy Physics,* Report 4 (New York: American Institute of Physics, 1992); Frederik Nebeker, "Experimental Style in High-Energy Physics: The Discovery of the Upsilon Particle" (unpublished), January 1993.

a 20-year period, which spanned three laboratory directors.[72] These continuities transcend a single experiment and provide a method for understanding more complex social structures and research programs that exist for more than 15 years. Each experimental configuration in a string displays a more complex iteration of the original apparatus that leaves the fundamental design of the modularized detector subsystems largely intact. In other words, experiment strings are like mini-institutions within the organizational infrastructure of the laboratory. People outside the laboratory really do not know about them because they do not have formal names.

The experiment string is the preferred and more interesting unit of study for sociological and historical analysis, because the numbers that laboratories such as Fermilab assign to experiments are not at all indicative of what actually constitutes "an experiment." Actually, the experimental numbers assigned by laboratory management are more indicative of such factors as the laboratory's accounting practices, the bureaucratic steps involved in the approval process as defined by a particular director, funding scenarios both inside and outside the laboratory, contrasts between the in-house/facility approach to doing experiments (where strings were largely determined by the laboratory management), and the user-based/non-facility approach (where institutions

72 The major fixed-target experiment strings at Fermilab were the E-82, 226, 383, 425, 486, 584, 617, 731, 773 string; the E-531, 653 string; the E-8, 440, 495, 555, 620, 619, 756, 800 string; the E-21A, 262, 320, 356, 616, 770 string; the E-594, 733 string; the E-98, 365, 665 string; the E-1A, 310 string; the E-95, 537, 705, 771 string; the E-70, 288, 494, 605, 608, 772, 789 string; the E-87, 358, 400, 401, 402, 687, 831 string; and the E-516, 691, 769, 791 string. By way of comparison with those counter experiments, the major continuity between the experiments performed with the 15-foot bubble chamber at Fermilab (experiments E-28A, 31A, 45A, 53A, 155, 172, 180, 202, 234, 341, 380, 388, 390, 545, 564, and 632) seems to be the chamber itself. In a less well-defined way, there were some continuities in the target substances with which the chamber was filled. But the social structures of these collaborations were different from the fixed-target counter experiments. Bubble chamber spokespersons seemed to draw upon the expertise of the international community of bubble chamber physicists each time they formed an experimental group, and consequently the collaborations did not exhibit the same type of well-defined core-group structure found in large, complex fixed-target counter experiments. My preliminary studies show that the relatively noncomplex social structure of these collaborations resulted from the existence of a Fermilab-based Bubble Chamber Department devoted solely to the operation and maintenance of the highly complicated systems of the chamber, independent of the collaborations that used it. This type of heterogeneous Fermilab/collaboration social structure with a dedicated support group is not evidenced in even the largest fixed-target counter experiments, but it is interesting to note that a similar phenomenon (dedicated support departments) does appear with the advent of the mammoth collider detectors such as CDF and D0. For historical details see Mark Bodnarczuk, ed., *Reflections on the Fifteen Foot Bubble Chamber at Fermilab* (Batavia: Fermilab, 1989).

came together and formed strings voluntarily). But these numbers do not define what an experiment is.

While it has been common practice for philosophers, historians, and sociologists of science to extract an experimental case study from the organizational infrastructure of the laboratory in which it was performed and study it as a stand-alone unit, Fermilab experiments did not exist independent of the organizational infrastructure in which they were embedded. Experiments and collaborations were not closed systems, cohesive entities, or "objects" that had unambiguous boundaries and could be divorced from the dynamics of laboratory life.

Of course, experimentalists did attempt to draw a firm line of demarcation around the "collaboration"—or "experiment"—and its activities for the sake of defining which names appear on scientific publications. But laboratory personnel often play crucial roles in experiments, and whether or not their names appear on the published paper is a socially negotiated matter decided by the people involved.[73] Attempts to "map" the names on various experimental proposals (or the resultant publications) to the collaboration members who actually performed the day-to-day tasks associated with the experiment show that the names on proposals or papers are often not indicative of those who actually performed the experimental work. Sometimes the names of people who played no substantive role in a given experiment are included on a proposal or on physics publications because they had major responsibility for constructing a portion of the detector in an earlier experiment in the string. Other times they may have committed a fraction of their overall professional time at the proposal stage but never came through on their commitments because of the heavy load of administrative duties at their home institution, the host laboratory, or their commitments to other experiments that they viewed as producing more important physics results.[74] This is related to

73 Melvin Schwartz shows how tenuous these socially negotiated walls are for today's large collaborations when he advocates divorcing some of the detector builders from the collaboration, then subdividing the remaining members of these megacollaborations into distinct (smaller) collaborations that would develop their own research programs and compete for time using the detector. In a sense, Schwartz is advocating a return to a social structure that is much like the one displayed in large bubble chambers—something I described in the previous note on the 15-foot bubble chamber at Fermilab. Also, see Faye Flam, "Big Physics Provokes Backlash," *Science 30* (11 September 1992), 1470.

74 Some collaborations (such as CDF) required members of the collaboration to run a certain

the commodity of "physicist economics," and may be a fruitful area for future sociological research.

Physicists' choices to go to higher and higher energies in response to the experimental evidence and theoretical constructs that emerged during the 1964 to 1979 period, and the effect of these choices on increasing scale, cost, and complexity, reveal interesting contrasts between the American and European (CERN) styles of doing physics. During this period many American physicists preferred the more informal, nonbureaucratic, quick-and-dirty, frugal style of doing physics.[75] But the European style was characterized by what American physicists saw as an overly formal, inflexible, bureaucratic, overengineered, "gold-plated" approach to doing physics. Even after the mammoth collider detectors began to be conceived in the late 1970s, both American and European physicists were relatively unreflective about the role that social factors were beginning to play in their work. And consequently, the sociological challenges that were intrinsic to collider detector environments with 10^5 or more channels received little or no systematic study by practicing physicists. The sociology of large collaborations just was not viewed as a part of doing high-energy physics and as with the policy of physics journals, the social and human factors were simply *left out.*[76]

---

number of data-taking shifts in order to have their name on publications, but many collaborations had no such policies.

75 In Chapter 20, Westfall refers to a similar type of nonbureaucratic, quick-and-dirty, frugal style at SLAC. Also see Catherine Westfall and Lillian Hoddeson, "Thinking Small in Big Science: The Founding of Fermilab — 1960–1972" in *Technology and Culture*, Johns Hopkins University Press, Vol. 37, No. 3, July 1996, 457-492.

76 Kevles attributes the scientist's tendency to leave social factors out of their accounts of science to being accustomed to the literary convention of journal editors and the fact that many scientists consider themselves to be incompetent to write about anything except the science itself. See Daniel Kevles, *The Physicists: The History of a Scientific Community in Modern America* (New York: Alfred Knopf, 1978), x. Pickering claims that references to "judgments" or "agency" on the part of scientists are left out of scientists' accounts so scientists can portray themselves as passive observers of nature, with experiments appearing to be the supreme arbiters of competing theories. Andrew Pickering, *Constructing Quarks: A Sociological History of Particle Physics* (Chicago, IL: The University of Chicago Press, 1999), 5-18. Latour claims there are definable processes that operate to remove all aspects of the social and historical context from scientists' accounts so scientific "facts" do not appear to be socially constructed. See Bruno Latour and Steve Woolgar, *Laboratory Life: The Construction of Scientific Facts* (Princeton, New Jersey: Princeton University Press, 1986), 176-183; and Bruno Latour, *Science in Action: How to Follow Scientists and Engineers Through Society* (Cambridge, MA: Harvard University Press, 1988), 22-29.

But despite this lack of self-reflection on both sides of the Atlantic, the values embodied in European culture more naturally gave rise to a style of physics that was more formal in terms of well-defined roles, responsibilities, and authorities for physicists and engineers. The European approach was also more focused on producing robust engineering and physics designs that were less flexible in terms of programmatic changes. As it turned out, these were the very practices, values, and beliefs that became crucial to mounting mammoth collider-detector experiments. Conversely, the less formal approach to doing physics put American physicists at a disadvantage in terms of confronting the kinds of organizational and management problems that emerged from this enormous growth in scale, cost, and complexity. While the American style of doing physics was an advantage with the scale, cost, and complexity of detectors in the 1964 to 1979 period, it became a crucial disadvantage for experiments conceived in the late 1970s, and was absolutely terminal for the proposed SDC and GEM detectors.[77] Also, the European style allowed a more natural transition from the smaller experimental scale that typified the 1964 to 1979 period to the detectors of the present day. In the modern detector environment, not only can social factors no longer be left out of any salient definition of what high energy physics is, but they become one of the most crucial aspects of doing high-energy physics—they may even become the limiting factor of the future of the field.

77 It is interesting to note that in an address in honor of the 75th Anniversary of the Max Planck Institute for Physics in Munich, Germany, James D. Bjorken devoted a major portion of his visionary article to the problems associated with the sociology of large collaborations and the possibility that these social factors might have an effect on the physics itself. See James D. Bjorken, "Particle Physics - Where Do We Go from Here?" SLAC *Beam Line,* Vol. 22, Winter 1992, 10.

The inside of Wilson Hall, the focal point of the Fermilab site, was designed by the laboratory's first director Robert Rathbun Wilson who was inspired by French Gothic architecture. The building is made of reinforced concrete, shaped by fiberglass forms cast from wooden boards which give the concrete walls their distinct texture. Two separate towers connected by crossovers soar high above an indoor atrium with seven-story glass walls that cast natural light on large fig trees and lush vegetation, creating an outdoor park-like atmosphere inside the building that Wilson envisioned as a "cathedral of science."

2

Megascience in Particle Physics: The Birth of an Experiment String at Fermilab

Mark Bodnarczuk and Lillian Hoddeson[78]

Abstract

During the 1970s and 1980s, the experimental particle physicists at large laboratories responded to the limited funding context of that time. Unlike the "big science" in the postwar decades, when funding for research appeared unlimited and many parameters grew exponentially, the even larger-scale "megascience" of the 1970s and on was shaped by competition for limited resources at host laboratories. Experimental programs increasingly took the form of long-lived (typically ten- to twenty-year) strings of related experiments, invisible institutions creating research traditions within the laboratory. Focusing on how a particular study of charmed particles (experiment E-516) at Fermi National Accelerator Laboratory gave rise to one of the laboratory's earliest experiment strings (E-516-691-769-791), we explore why such strings evolved and how they led to new research practices. Because these strings produced conflicts and ironies that threatened to undermine fundamental aspects of the research, the emergence of experiment strings can be viewed as a limiting process of large-scale research in the 1970s and 1980s.

The term "big science" came into use after several writers used it in the early 1960s to refer to what at the time appeared a boundless growth of research.[79] Many historians of science have since examined aspects of big

78 This article was originally published in *Historical Studies in the Natural Sciences,* Vol. 38, Number 4, 508–534. ISSN 1939-1811, electronic ISSN 1939-182X.

79 Classic references include Derek J. De Solla Price, *Little Science, Big Science* (New York: Columbia

science, typically attributing the phenomenon to new government funding sources created in the United States during or shortly after the Second World War that channeled bountiful support into many areas of science. Particle physics grew very rapidly in the first two postwar decades, when this field was entering its scientific adulthood. As many historians of science (including Peter Galison, Daniel Kevles, Robert Seidel, Silvan Schweber, and Peter Westwick) have argued, the field benefited in those early Cold War years from the association of military advantage with the expertise of the corps of highly trained particle physicists. Even today, much of the writing about big science in particle physics refers to the first two-and-a-half postwar decades.

Much less examined is the big science seen in the decades following 1970 when funding limitations shaped research in new ways. The trends were notable in particle physics, where the size, complexity, and cost of most experiments continued to grow even as researchers found themselves facing stiffer competition for resources. The struggle for high-energy physics experiments to grow in a limited funding environment affected the physicists' scientific practices in paradoxical

University Press, 1963); Alvin M. Weinberg "Impact of Large-Scale Science," *Science* 134 (1961), 161–64; and Alvin M. Weinberg, *Reflections on Big Science* (Cambridge, MA: MIT Press, 1967). Writers have used the term in many fields, from genomics to weapons research, whenever that field's parameters (e.g., size, cost, time spans, scientific intricacy, complexity of apparatus, or numbers of scientists) appear large. However, the term is most often applied to the early Cold War decades, when such parameters appeared to be increasing exponentially. For the present view on big science and many references, see Catherine Westfall, "Rethinking Big Science: Modest, Mezzo, Ground Science, and the Development of the Bevalac, 1971–1993," *Isis* 94 (2003): 30–56; Peter Galison and Bruce Hevly, eds., *Big Science: The Growth of Large-Scale Research* (Palo Alto, CA: Stanford University Press, 1992); James H. Capshew and Karen H. Rader, "Big Science: Price to the Present," *Osiris* 7 (1992): 3–25; Wesley Shrum, Joel Genuth, and Ivan Chompalov, *Structures of Scientific Collaboration* (Cambridge, MA: MIT Press, 2007), 1–7. Historians do not agree when big science began. Historians of technology often point to the big science done circa 1900 in early industrial laboratories, such as General Electric or Du Pont, e.g., George Wise, *Willis R. Whitney, General Electric and the Origins of U.S. Industrial Research* (New York: Columbia University Press, 1985) or David A. Hounshell and John Kenly Smith, Jr., *Science and Corporate Strategy: Du Pont R&D, 1902–1980* (Cambridge: Cambridge University Press, 1988). In particle physics it is accepted that big science emerged in nuclear physics research in Berkeley during the 1930s at the time of Ernest Lawrence's pioneering cyclotrons: J. L. Heilbron and Robert Seidel, *Lawrence and His Laboratory: A History of the Lawrence Berkeley Laboratory*, vol. 1 (Berkeley: University of California Press, 1989). Weapons historians look later at laboratories organized during or just after the Second World War: L. Hoddeson, P. Henriksen, R. Meade, and C. Westfall, *Critical Assembly: A History of Los Alamos during the Oppenheimer Years, 1943–1945* (New York: Cambridge University Press, 1993). Historians of space science may see the late 1950s or 1960s as the creation years for that field: Robert W. Smith, Paul A. Hanle, Robert H. Kargon, and Joseph N. Tatarewicz, *The Space Telescope: A Study in Science, Technology, and Politics* (Cambridge: Cambridge University Press, 1993).

ways, as we will discuss below. Their day-to-day activities changed so much that we refer to the bigger big science of the 1970s and 1980s with a new name, "megascience." This term has previously been used by other writers to simply mean very big science.[80] We use it in a more restricted sense, to refer to the changed big science seen from 1970 on, when funding no longer appeared limitless.

As the name implies, megascience is in some ways even bigger than the earlier big science—in the size of its groups, budgets, bureaucracy, timescales, and data sets.[81] But there is a crucial sense, as we argue, in which this megascience is actually smaller than the big science of the postwar decades. As the physicists worked to make their precious resources go farther, especially by conducting follow-up studies in which aspects of the work were reused or extended, their experiments were often connected with each other in strings of related studies whose continuities (e.g., in the physics or their apparatus) reflected long-lived (typically fifteen- to twenty-year) traditions.[82] The subsequent experiments in a string were in some sense iterations of the original experiment, differing from it only in limited ways, e.g., a new aspect of the research problem, a modification of the apparatus, a new computing architecture, a new approach, or new members of the collaboration. Forming invisible

80 See for example, Jeff Hughes, *The Manhattan Project: Big Science and the Atom Bomb* (New York: Columbia University Press, 2002).

81 The growth in scale of detectors in the late 1970s produced an explosion in the number of electronic channels in the particle physics detectors. The UA1 detector at CERN, for example, had some 50,000 channels, while the CDF and D0 detectors at Fermilab and the SLD detector at SLAC each had over 100,000 channels. The ALEPH detector at the Large Electron-Positron Collider (LEP) at CERN had about 700,000 channels, while the SDC and GEM detectors proposed for the discontinued Superconducting Super Collider would have had as many as 50,000,000 channels.

82 One of us (Bodnarczuk) pointed out the social structure that we refer to here as an experiment string in a March 1990 Fermilab internal publication, "The Social Structure of Experimental Strings at Fermilab; A Physics and Detector Driven Model," *Fermilab-Pub-91/63*, FA. See also Bodnarczuk, "Some Sociological Consequences of High-Energy Physicists' Development of the Standard Model," in *The Rise of the Standard Model: Particle Physics in the 1960s and 70s*, ed. Lillian Hoddeson, Laurie M. Brown, Michael Riordan, and Max Dresden (New York: Cambridge University Press, 1997), 384–393. A few other scholars subsequently picked up this notion in the course of studying multi-institutional collaborations at Fermilab and other laboratories, e.g., Joel Genuth, "Historical Analysis of Selected Experiments at U.S. Sites," *AIP Study of Multi-Institutional Collaborations: Phase I: High Energy Physics, Report 4* (New York: American Institute of Physics Press, 1992); Frederik Nebeker, "Strings of Experiments in High-Energy Physics: The Upsilon Experiment," *Historical Studies in the Physical and Biological Sciences* 25, no. 1 (1994): 137–64. For more about experiment strings at Fermilab, see Lillian Hoddeson, Adrienne Kolb, and Catherine Westfall, *Fermilab: Physics, the Frontier, and Megascience* (Chicago: University of Chicago Press, 2008), esp. chaps. 1, 11, and 12.

mini-institutions within the programs of large physics laboratories, such strings are among the most striking characteristics of megascience. Strings formed invisible mini-institutions within the laboratory.

Experiment strings can hardly be avoided in research areas that employ a large or expensive apparatus, such as a colliding-beam detector or a space telescope, which is intended to last at least for a generation. But when resources are limited, even experiments that in principle could be disassembled at the end of an experiment run tend to be extended through the formation of a string. At Fermi National Accelerator Laboratory (Fermilab), in Batavia, Illinois, the experimental program of the 1970s to the 1990s contained approximately a dozen experiment strings (the number depending on how one counts), most of them centered on expensive and complex detectors. By studying how in the early 1970s one large Fermilab physics experiment gave rise to one of the earliest experiment strings there, we seek insight into why such strings emerged so often at large laboratories after 1970 and how this phenomenon threatens to undermine certain fundamental aspects of large-scale research in high-energy particle physics.

The Fermilab experiment on which we focus, E-516, was a pioneering study of charmed particles (sometimes called "charm particles")—hadrons containing at least one "charm" quark, one of the six kinds of quarks in the theory of elementary particles known as the Standard Model.[83] This experiment along with its three follow-up experiments, E-691, E-769, and E-791, comprised a string, which, from the conception of E-516 through E-791's data analysis, spanned two decades (1971–92). In all four experiments, the beam of accelerated protons emerging from the laboratory's accelerator first struck a primary production target located hundreds of feet upstream of the experiment hall known as the Tagged Photon Laboratory (TPL), the "stream" beginning at the accelerator, continuing on "down" through the TPL and into the experimental target and the detector, known as the Tagged Photon Magnetic Spectrometer (TPMS).

The impact of the protons on the primary production target created a secondary beam of particles, some of which were electrons. These electrons subsequently penetrated a thin metal foil just upstream of the TPL, producing a photon beam in the case of E-516 and E-691, and a hadron beam in

83 "Charm" quarks and "strange" quarks are in the middle generation of the particle physics theory known as the Standard Model. The other four quarks are: "up" and "down" quarks in the first generation; and "top" and "bottom" quarks in the third generation. For further discussion, see Hoddeson et al., *Standard Model* (ref. 4), 3–31.

the subsequent two experiments. Because the energy of the electron beam was measured ("tagged") with a particular energy for studying particle interactions of interest, the photon or hadron beam was also "tagged." Finally, a beam of these created photons or hadrons impinged on yet another target to produce the "charmed states" which were examined in the TPMS. The TPL and the TPMS were costly pieces of real estate; such experimental real estate will emerge as a central actor in this account of strings and megascience.

The Birth of Experiment 516

Our story of the birth of the E-516 string begins in the summer of 1971 at the National Accelerator Laboratory (NAL, later renamed Fermilab in 1974). Robert R. Wilson, Fermilab's founding director, was in the last stages of building the "Main Ring," the laboratory's original accelerator. Anticipating the start of experimental work using protons accelerated to high energy by the Main Ring, three groups of physicists submitted competing proposals to create the TPL. The laboratory assigned numbers to the three proposals, all prefaced by the letter "P" to indicate their proposal status. All three of the proposed experiments planned to examine one or more aspects of the photons to be produced in the TPL: P-25 proposed studying photon total cross sections; P-144 aimed to study general features of photon-induced reactions in the 50 to 500 GeV region; and P-152 proposed studying production of the photon (i.e., photoproduction).

The influence of limited resources and a frugal philosophy concerning their use at NAL was immediately evident when NAL's management explained to the proposing groups that for budgetary reasons their three efforts would have to collaborate on a single TPL design that met the needs of all three teams. Samuel Ting, the spokesperson for P-144, promptly withdrew his proposal. The competition continued between P-25 and P-152, represented by their spokespersons, David Caldwell and Clem Heusch, respectively. When the laboratory approved both proposals, they became E-25 and E-152. Both now planned to "run" their experiments in the beam during a specified number of hours. When E-25 committed itself to building the beamline, the pipe in which the particle beam travels to their experiment, this group was awarded first use of the TPL while E-152 was awarded second use.

The experimental physicist E. Thomas Nash had worked at DESY, Hamburg's high-energy physics laboratory, with Ting, who sent Nash to

NAL to help design P-144. When Ting withdrew P-144 from the TPL competition, Nash was out of a job. He decided to join NAL and become a member of E-25. While that experiment ran from 1971 to 1976, he proceeded to organize a new collaboration, which in time became E-516. Besides Nash, who would serve as spokesperson, the new group included George Luste from the University of Toronto, Rolly Morrison from the University of California at Santa Barbara, Jeff Appel from Columbia University, and Paul Mantsch from Fermilab. This team set as its initial goal to build a sophisticated spectrometer that could fully utilize the tagged photon beam in the TPL to study the production of charmed particles. This detector, the TPMS, would be the centerpiece not only of E-516, but of all three of this experiment's follow-up studies. To produce the charmed particles, Nash's team planned to shine a photon beam on a liquid hydrogen target. The computing architecture of the TPMS would let it select, record, reconstruct, and analyze charmed events. The aim was to elucidate such properties of charmed particles as production, cross section, and mass. By early June 1976, the new team was ready to submit its "Letter of Intent" to the Fermilab directorate proposing the experiment.[84]

Even at the start, funding limitations influenced the experiment. To optimize its chances of receiving running time, Nash's group aligned itself with Wilson's well-known commitment to frugality in all aspects of the laboratory's activities.[85] The frugality extended to the use of computers, an essential data-gathering and analysis resource for any experiment employing a complex detector. Wilson ruled that NAL should offer experiments minimal computing facilities, not only "online" computing support, where the computers are physically connected to the experimental apparatus with cables, but "offline" computing support, where the computers are not attached to the apparatus and are located elsewhere at the laboratory, or at another institution. For the scientists who were visiting the laboratory from other institutions, Wilson was adamant that the offline computing support for these "users" should come from their home institutions. Respecting Wilson's limitation on computing support, the E-516 team planned not to take the typical "write-it-all-to-tape" approach, which gathered the most data but required more offline computing

84 Thomas Nash to R. R. Wilson, E. L. Goldwasser, and the Fermilab Physics Advisory Committee (PAC), 6 Jun 1976, NP.

85 For a full explanation of Wilson's frugality philosophy, see Catherine Westfall and Lillian Hoddeson, "Thinking Small in Big Science," *Technology and Culture* 37, no. 3 (1996): 457–92.

support than the experiment expected. Instead, the group planned to reduce its data using an electronic trigger processor designed by Nash and Steve Bracker of Toronto. This device would turn the online recording system on only when the detector identified the signature of a charmed particle. Determined by the best available theory, the "diffraction model," this signature was a recoil proton moving at a certain angle with a particular momentum.[86] The reduced data recorded would then be sent to offline computer programs, which would "reconstruct" the charmed particles.

On receiving the Nash group's "Proposal to Study Photoproduction of Final States of Mass Above 2.5 GeV with a Magnet Spectrometer in the Tagged Photon Lab," the Fermilab directorate assigned number 516 to the proposal, and it then became known as P-516. The laboratory also agreed to give this experiment, if approved, first use of its new spectrometer. With its newly acquired status as a numbered Fermilab proposal, the P-516 collaboration could now recruit additional physicists to help plan the experiment and design its TPMS.

Approval for E-516

By the standards of its day, the TPMS was an unusually complex particle detector. Designed to detect eight to twelve charmed particle decay products, it had some 7,000 points where particles would interact. As shown in Figure 1, these were distributed over six detection regions of the TPMS: (1) the recoil detector; (2) drift chambers; (3) Cerenkov counters; (4) electromagnetic calorimeters; (5) hadron calorimeter; and (6) the muon wall.[87] The recoil detector was used with the experimental target to trigger the spectrometer; the drift chambers were used to track charged particles by analyzing their trajectories and momenta; the Cerenkov counters were used to identify particle types by measuring their mass; the electromagnetic calorimeters, consisting of the segmented liquid ionization calorimeter (SLIC) and outriggers, measured

86 Before quantum chromodynamics (QCD), the diffraction model was the prevailing theory. It was based on the framework of "vector dominance" championed in the 1960s by Jun John Sakurai. This theory specified that the strong interaction between particles could be explained in terms of the exchange of vector mesons. See the proposal for P-516, "Proposal to Study Photoproduction of Final States of Mass Above 2.5 GeV with a Magnet Spectrometer in the Tagged Photon Lab," 11, and the accompanying footnote. For more details of the initial beam design, see P. Davis, R. Morrison, T. Nash, J. Prentice, J. Cumalat, R. Egloff, G. Luste, and F. Murphy, "The Electron Beam Test of October–November 1974," *Fermilab TM -535* (6 Dec 1974), FA.

87 The TPMS spectrometer's design is discussed in J. Appel, D. Bartlett, S. Bracker et al., *The Tagged Photon Magnetic Spectrometer: Facility Design Report*, 9 May 1977, NP.

the energy of photons and electrons; the hadron calorimeter measured the energy of uncharged hadrons; and the muon wall detected the presence and measured the trajectories of muons that penetrated the forward region of the spectrometer. The six regions, designed to function together as a unit, were each connected to channels of electronics that fed into the trigger processor and then into the online computing system.

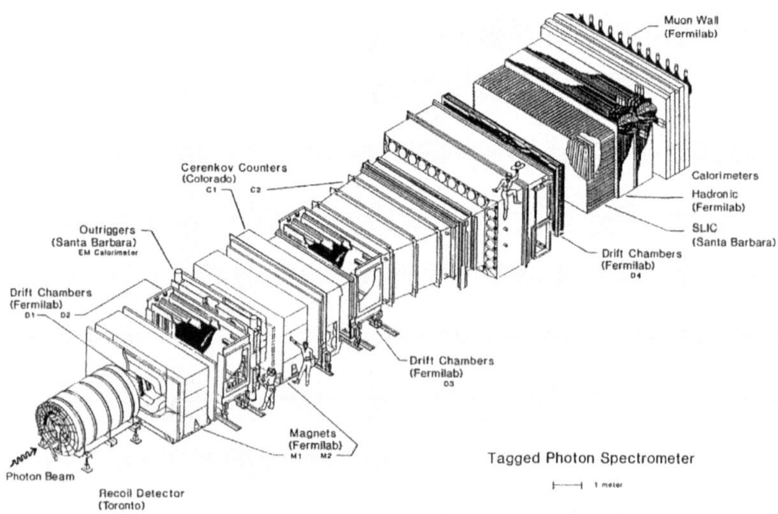

Figure 1: The Tagged Photon Magnetic Spectrometer.

Because each region of the TPMS was built and maintained by one or more of the collaboration's member groups, the design of the detector shaped the collaboration's social structure, forming a physical organization chart. The University of Toronto group was in charge of the experimental target and recoil detector; Toronto and Fermilab collaborated on the trigger processor; the Santa Barbara group was responsible for the electromagnetic calorimeter; the University of Colorado took charge of the Cerenkov counters; and Fermilab assumed the responsibility for the drift chambers, hadron calorimeter, and muon wall.

In presenting the proposal of their experiment, the P-516 collaboration took Wilson's frugality seriously, portraying the TPMS as a device built partly out of apparatus recycled from previous experiments. The collaboration planned to later ratchet up the laboratory's commitment of resources, including magnets, computing power, and beam time. Wilson likened this strategy for extracting resources to a "camel trying to get its nose under the tent."[88] In an effort to deemphasize the expense of the TPMS to the laboratory, its design report argued that it would be a general-purpose facility intended for use in a variety of experiments by different groups. In practice this claim proved empty, for E-516 and its three follow-up experiments would hold on to this costly piece of experimental real estate, and the TPL, throughout the duration of the experiment string (except for one interlude when E-612 ran parasitically during E-516's scheduled operation).[89]

Wilson also insisted that his laboratory, including the experiments conducted there, meet the needs of the outside users. That was why he had pressed to have the laboratory named "*The* National Accelerator Laboratory" in the first place. He wanted to distinguish NAL from laboratories, such as Lawrence-Berkeley or Brookhaven, where users had complained that their needs were secondary to those of staff physicists. In line with Wilson's goal to make his laboratory respond to its users' interests, P-516 made sure that "in-house" NAL groups did not control the collaboration and that responsibilities were shared between in-house and outside user groups.[90] The outside users would benefit from their prominent status in the experiment back at their home universities, where the description of their responsibilities at NAL would contribute to the basis for determining their promotions and salaries.

NAL's powerful Program Advisory Committee (PAC) was the body of experts advising the director of the laboratory on whether an experiment deserved running time in the laboratory's proton beam, and, if so, how much time and with what level of funding. This committee met twice each year, in June and November, to review the submitted experiment proposals, evaluating their importance, their plan, and how well their plan fit the available resources. The many considerations addressed by the committee included, for example, whether a collaboration involved enough people having the training and experience needed to bring the experiment to success.[91] Despite P-516's strenuous efforts to design an experiment suitable for running at NAL, Wilson's PAC did not consider P-516 ready for approval in November 1976, and he deferred

88 Robert R. Wilson, interview by Mark Bodnarczuk, 24 Sep 1992, FA.

89 Thomas Nash, interview by Mark Bodnarczuk, 18 Jul 1990, FA.

90 For more discussion of Wilson's democratic ideals see Hoddeson et al., *Fermilab* (ref. 4), chap. 5.

91 Edwin L. Goldwasser to Thomas Nash, 19 Nov 1976, NP.

the experiment. Recognizing, however, that once P-516 was approved, its team would need more time to build the TPMS than other groups typically needed to build their detectors, Wilson advised the collaboration to go ahead and design the TPMS while preparing to resubmit P-516 to the PAC the following June. By now the team consisted of fourteen physicists from four institutions.

During their long wait to run, P-516's members built prototypes for components of the TPMS. They waited longer than expected because in the tight funding environment of mid-1977 Wilson had to defer P-516 again after the June 1977 PAC meeting. The size and complexity of the TPMS simply pushed the limits of the laboratory's resources, no matter that the group had described its experiment as frugally designed. The caution with which Fermilab proceeded with its approval of P-516 was further illustrated when the collaboration requested another $1 million that month to build new analysis magnets. Wilson and his deputy director Edwin Goldwasser asked that they first build prototypes and offer "a few specific examples of the kinds of measurements which you anticipate making during the initial operating period of the facility." The laboratory wanted evidence that the collaboration had the expertise on its team to succeed, and it also was concerned that P-516 might have underestimated the cost of its project in the interest of being approved in the prevailing tight funding context.[92]

Considerable jockeying took place during the group's long wait to resubmit its proposal. It was not possible for the P-516 group to set up and test its apparatus in the TPL because E-25 was still running there. When E-25 completed its run, the next experiment in line, E-152, moved in. E-152 was a University of California at Santa Cruz group, which had been promised only one run in the TPL. It was especially vulnerable to displacement because it was the last modern high-energy physics experiment to be mounted by a single institution. Clem Heusch, the experiment's spokesperson, recognized E-516 as a prodigious threat, for once the enormous TPMS had been installed in the TPL the larger experiment would be unlikely to leave the TPL soon. He therefore proceeded to oppose the displacement by angling for additional time and the possibility for a follow-up experiment for E-152. Early in June 1977, he informed NAL's directorate that he planned to request additional running time on completing his run in the TPL. Anticipating Wilson's question of what measurements were planned with the additional running time,

92 Edwin L. Goldwasser to Thomas Nash, 29 Jun 1977, NP.

Heusch said he could not define his physics goals until completing his initial run. He played coyly on the laboratory's (and P-516's) promise to make the TPMS available to other users, claiming "that as much equipment as is technically feasible would be shared among different efforts."[93]

P-516 was threatened too. If Heusch were granted his request before the end of his first run, P-516 would be delayed for several years. Nash quickly offered technical reasons why E-152's detectors would not meet P-516's needs.[94] But Heusch won the round. Fermilab approved Heusch's request, granting 1,800 hours of additional beam time to E-152. Such maneuvering for the chance to run in the beam of a large accelerator would be characteristic of the new megascience.

By October 3, 1977, Nash had addressed all the issues raised by the PAC and the Fermilab directorate.[95] P-516 was finally approved in November 1977, more than six years after the experiment had been conceived.[96] In approving E-516, Wilson apportioned 1,000 hours of beam time to the collaboration. The first 400 were specified for an initial shakedown and engineering run during the summer of 1979 for testing all the apparatus and showing that it would be possible to deliver the design flux of 10^6 photons to the experiment target set about a mile from the accelerator. The second run, the first to gather actual data, would begin in November 1979 and continue into early 1980. The third and final run would start in October 1980 and continue until the accelerator was shut down in June 1981. The elaborate process of reconstructing the particle events seen in the TPMS would take yet another year and a half. This long process, which all large experiments suffered in gaining approval to run at NAL, caused Wilson to break from his earlier policy of approving numerous small "quick and dirty" experiments rather than a few larger experiments of longer duration. Over time, this change helped the laboratory compete more successfully with other laboratories that more readily funded larger experiments.[97] As for Heusch and E-152, Wilson's approval of E-516 was bad news. E-152 lost its place in the TPL.

---

93 Clem Heusch to Edwin L. Goldwasser, 2 Jun 1977, NP.

94 Thomas Nash to Edwin L. Goldwasser, 15 Jun 1977, NP.

95 Thomas Nash to Edwin L. Goldwasser, 3 Oct 1977, NP.

96 Edwin L. Goldwasser to Thomas Nash, 15 Nov 1977, NP.

97 Westfall, "Thinking Small" (ref. 7).

E-516's Computing Crisis

Employing the trigger processor meant that the E-516 group would gather less data and be in a position to adhere to Wilson's frugality in computer support. The team therefore requested only a modest computer, a PDP11-55, for online processing of the acquisition and monitoring of the experiment's data.[98] But in the weeks following the approval of E-516, Steve Bracker, who had worked with Nash on the design of the trigger processor, began worrying that the requested minimal computer might not be adequate for the experiment. Nash pressed John Peoples, the head of Fermilab's Research Division, for more computing resources, and Peoples passed the request on to Al Brenner, the long-time head of the Computing Department, who controlled the computing resources offered to experimenters. Not only did Brenner support Wilson's philosophy of frugality in computing, he also understood, as did Peoples, that it was typical for experiments with large computing needs to appeal to frugality in their proposals but then request more support after approval.[99] Brenner upgraded E-516's online computer from the PDP11-55 to a PDP11-55T, but the upgrade amounted only to adding a tape drive; it did not increase the processing power. That limitation distressed some members of the collaboration. In late December 1979, two senior members, Morrison and Uriel Nauenberg of Colorado, warned Leon M. Lederman, who on June 1, 1979, had become Fermilab's second director, that E-516 was in danger of failing because of its inadequate computer support.[100] No action was taken, however.

The computing issue would blow up to crisis proportions during the second half of 1980, after the end of E-516's second run. By this time the level of anxiety within the collaboration was elevated because E-516 had already used two-thirds of its allotment of beam time but had not recorded any charmed particles. The collaboration members had not expected immediate evidence of charmed particles, but they did not imagine they would not see *any* such particles. Where were the charmed particles that physicists at the Stanford Linear Accelerator Center and at Brookhaven National Laboratory saw in their detectors? E-516's failure to detect charmed particles during its first two runs was a grave problem that deserved extended discussion by members of the collaboration. But in the social setting of the high-energy physics laboratory,

98 "The Online System for Experiment E-516: How Big Is It?," 11 May 1978, NP

99 Richard Carrigan, interview by Mark Bodnarczuk, 18 Aug 1992, FA.

100 Rolly Morrison and Uriel Nauenberg to Leon L. Lederman, 26 Dec 1979, NP.

it was not prudent to discuss such a matter publicly, not even at collaboration meetings. The approval of additional funds is based largely on performance, and competing experiments always lie waiting to displace experiments that are not making good on their promises. In July 1980, Nash wrote a memo warning the collaboration about the dangers of "loose talk." "Credibility is destroyed by the public statement of incorrect or, more commonly, variable results," he explained. "It takes judgment and a thorough, open and continuing discussion within the group of any work to determine when a result has stabilized and is dependable. Until this point is reached it is essential that results—and difficulties in the analysis—not be discussed idly." Nash also reminded his team that the "funding and beam time . . . are in such short supply that administrators are looking for any excuse to cut back resources—even if already granted. Under these conditions, a slight comment about [trouble with] a couple of bad channels in D1 returns as 'Their drift chambers don't even work. What do they need all that intensity for?'"[101]

The physicists who had worked at NAL during the early years of its experimental program recalled the embarrassment that many at the laboratory felt when, in an episode referred to as "alternating currents," the results reported from E-1A, the neutral current experiment of David Cline, Alfred Mann, and Carlo Rubbia, varied publicly. E-1A was competing with CERN to discover neutral currents and had claimed this major discovery, but it subsequently retracted this claim, stating that neutral currents did not exist, only to issue yet a second retraction, announcing that neutral currents did exist after all.[102] The episode taught experimenters at Fermilab a hard lesson: let results stabilize before announcing them. At the same time, however, gaining additional resources was contingent on publishing results.

By early December 1980, E-516 saw itself "in a very real crisis," as stated in correspondence exchanged between members of the collaboration. On December 9, seven weeks into the experiment's final run, Nash sent Morrison a letter expressing his concern that "there is more than a zero chance that we

---

101 Thomas Nash to E-516 collaboration, 21 Jul 1980, NP.

102 Peter Galison has highlighted the move from table-top-size spark chambers to the mammoth spark chambers with which Fermilab's E-1A searched for neutral currents; Galison, "How the First Neutral Current Experiments Ended," *Reviews of Modern Physics* 55 (1983): 477–509. See also Donald Perkins, "Gargamelle and the Discovery of Neutral Currents," in Hoddeson et al., *Standard Model* (ref. 4), 428–46.

will come up empty-handed."[103] This letter followed more than five years of preparation and an immense financial investment. The pressure on E-516 to announce results heightened when Lederman set February 1, 1981, as the deadline for submitting experiment proposals for the laboratory's higher-energy Tevatron program, based on its new superconducting accelerator, known in that period as the Energy Doubler. This program was projected to begin in two years.[104] In the midst of what appeared to be a failing experiment, Nash and his collaborators had less than two months to design a follow-up experiment to E-516 or lose its place in the TPL. Thus in the same letter to Morrison, in which Nash described the crisis of E-516, he added: "I would like to assure you that I (with your and others' help and advice) can put together a respectable document that will serve to hold our place in line."[105] Proposing a follow-up experiment before completing one was another common feature of the new megascience; it was a way to hold on to experimental real estate at a large laboratory. But such a proposal is usually made only when a collaboration has data to support its request. E-516 lacked such data, still the group could not afford to lose its "place in line."

By this time the cause of E-516's problems was becoming clear to some members of the collaboration.[106] Charmed particles were in fact penetrating the TPMS, but they were not being recorded to the experiment's magnetic data tapes, because E-516's triggering specification of a recoil proton was based on an earlier theory, the diffraction model. Within the framework of the theory that succeeded the diffraction model, quantum chromodynamics (QCD), the signature of a recoil proton would not dominate.[107] With this understanding, Morrison, Appel, and others in the group proposed running the TPMS with a less-biased trigger, and they pushed to eliminate E-516's recoil detector with its tight triggering parameters from their follow-up experiment proposal. If adequate offline computing were available, the team would be able to detect

---

103 Thomas Nash to Rolly Morrison, 9 Dec 1980, NP.

104 Norman Gelfand to Thomas Nash, 25 Sep 1980, NP.

105 Thomas Nash to Rolly Morrison, 9 Dec 1980, NP.

106 Rolly Morrison, interview by Mark Bodnarczuk, 3 Jul 1990, FA,

107 Thomas Nash to Edwin L. Goldwasser, 3 Oct 1977, NP. The evidence for this diffraction model came from experiments performed at energies below 60 GeV, but the physics was different when the beam came from the Fermilab Tevatron, whose 800 GeV proton beam caused the photon beam to have an energy of about 300 GeV.

the as-yet-unseen charmed particles using solid-state detectors placed directly downstream of the experimental target. Nash, however, having invested years developing the recoil detector and its trigger parameters, was not ready to give these up. When he argued to retain them in the proposal for the follow-up experiment, the dissenting members found themselves in an uncomfortable position, because they opposed the collaboration's most powerful member, its spokesperson, the one with the political power to overrule group consensus. During January 1981, the circulating drafts of the collaboration's proposal for its follow-up experiment reflected the group's fragmentation over its experiment plan.

By early February, the team had worked out a compromise in which the TPMS would retain the recoil detector and trigger processor, but the proposal would also discuss QCD and include a plan to study both diffractive and non-diffractive photoproduction of charm.[108] The directorate then assigned number 691 to this revised proposal for a follow-up to E-516 and requested the preparation of a statement of impact to help determine how much new equipment and computing resources would be needed and what effect the experiment would have on the resources of other proposed Tevatron experiments.[109] The assignment of an official proposal number to E-516's first follow-up experiment marked the birth of an experiment string, among the first such strings in Fermilab's experimental program in that period. The collaboration remained in an awkward position, however, because of its division over whether to retain the recoil detector with its tight triggering assumptions in the design of P-691. Morrison, Appel, and an increasing number of the others were opposed while Nash was in favor. The members argued about this issue for months.

The Reconstruction Shootout

After an experiment runs at a high-energy physics laboratory, the focus typically shifts toward data analysis. For E-516 and its follow-up experiments, this analysis involved reviewing the tracking system software, which "reconstructs" the particle events from the data gathered by the detector. In June 1981, when

108 Jeff Appel to Thomas Nash, 30 Jan 1981; and Thomas Nash to Norman Gelfand, 4 Feb 1981, both in NP, which includes a description of the proposal submitted to Fermilab's directorate and designated as P-691.

109 Norman Gelfand to Thomas Nash, 12 Mar 1981, NP.

Nash reported to the PAC on the status of E-516, he had to admit that the experiment's tracking system reconstruction programs were not ready. As an experienced experimenter, Lederman understood instantly that the experiment had not even begun its in-depth data analysis and was therefore far from publication. After the meeting, Lederman informed Nash that he was going to defer P-691 and would not grant approval until he received a clear statement of E-516's results.

E-516's crisis over its lack of data was compounded by a social storm that arose in the collaboration during 1981 over the experiment's particle reconstruction software. For over two years, Uriel Nauenberg and John Bronstein, two senior members of the collaboration, had been writing independent particle reconstruction software packages for the experiment's tracking system. While high-energy physics experiments rarely find the resources to build two detector subsystems that perform identical functions, the collaborators often write more than one software package for a subsystem. Such programs reflect the styles of their authors; multiple versions can cross-check each other. Neither Nauenberg nor Bronstein knew, however, that a junior member of the collaboration, David Bintinger, had been quietly working on a third software reconstruction package for the tracking system. Bintinger had not wanted to offend his senior colleagues, but as he neared completion of his package, he found himself in a delicate situation. He showed his work to Nash, who appreciated its merit and who, as the spokesperson, had to decide how to present Bintinger's effort to the rest of the collaboration. Nash planned how to compare the quality and efficiency of all three programs. But before Nash could complete his planning, the news of Bintinger's program leaked.

As expected, Nauenberg and Bronstein were disturbed that Bintinger had moved in on their territory, and that Nash apparently supported Bintinger. Nash responded to Nauenberg and Bronstein in a formal letter. "David's right to pursue the direction of his interests is, of course, the same right you both had when you started your reconstruction efforts." Nash proposed settling the issue with a "shoot-out."[110] He asked Bracker, who had substantial expertise in computing hardware and software and was not involved in the dispute, to serve as referee. Bracker refused to examine any of the programs before he had developed quantifiable criteria for comparing them. In an effort to eliminate infighting over the definition of the criteria, he recorded his daily

110 Thomas Nash to Uriel Nauenberg and John Bronstein, 24 Sep 1981, NP.

progress on the task in a computer file accessible to all members of the collaboration. Eventually, all agreed that Bracker had drawn up an acceptable basis for comparing the three competing software packages. At the shoot-out, the three programs were run and evaluated with identical data sets. When the smoke cleared in November 1981, only Bintinger's program was left standing.[111]

The E-516/691 Interface and the Computing Revolution

To everyone's relief, in the early months of 1982, owing to the tuning of existing hardware and software, as well as help from Bintinger's new reconstruction code, the experiment began to show preliminary evidence for charmed particle reactions. The evidence came at about the same time that the laboratory notified E-516 that it would have to submit its annual report. The collaboration was able to claim that reconstruction and analysis of the results from E-516 yielded "encouragement in the charm channels that are our primary interest." And it could also submit the E-516 program report in requesting approval for P-691. It was nevertheless apparent on May 6, 1982, when the team requested approval of P-691, that after a year of data analysis, E-516 had managed to reconstruct only some 15 percent of its total data sample of 20 million triggered events.[112] Insufficient computer support deriving partly from Wilson's policy of frugality was largely to blame for E-516's inadequate data. Had there been enough offline computer support, the experiment could have avoided working with a faulty triggering assumption. Similarly, with regard to the P-691 proposal, Morrison and Appel's suggestion to run the spectrometer with a less-biased trigger could be implemented only if low-cost, high-power computing were available—computing that allowed all the evidence for the charmed particles entering the spectrometer to be written to magnetic tape. Lederman recognized the importance of improving Fermilab's computing facilities, but such correcting of the situation would prove to be a time-consuming process.

The discussions that now ensued about E-516's inadequate computer support illustrate how the problems of a single almost failing experiment can bring major change to a large laboratory. Nash took the lead. Because of his involvement with the trigger processor for E-516, he was among the Fermilab physicists most disabled by the computing limitations. He devoted much thought to how best to correct this problem. On May 11, 1982, five days

111 Steve Bracker, "Bracker's New Track News," 3 Nov 1981, NP.

112 Thomas Nash to Taiji Yamanouchi and Norman Gelfand, 6 May 1982, NP.

after submitting the E-516 progress report, Nash submitted another proposal to Fermilab, "A Program for Advanced Electronics Projects at Fermilab," explaining how experiments grow naturally until they reach a point at which their data is limited by "the amount of computing time they anticipate will be possible to squeeze out of the system." Clearly, the more data an experiment is able to process, the more likely it is to succeed. Without referring to E-516, Nash explained the collaboration's difficulties in obtaining results in terms of the computing problems confronting all large Fermilab experiments. In almost a confession of E-516's near failure to detect the presence of charmed particles, he argued that E-516 was "up against the technological barrier of computing power limitations."[113]

As discussed by Robert Seidel in this issue of *HSNS*, Nash went on to propose a new program by which Fermilab could address the data problems of large physics experiments. The program focused on the development of multiple parallel-processing computing systems based on powerful, specially configured Intel 32-bit microprocessors. He explained that such systems, if properly configured, would offer experiments a thousand-fold increase in cost-effectiveness, for the reduction in cost and increase in computing power would allow experiments to use less-biased trigger assumptions and record more data to tape while at the same time accelerating the data analysis leading to publication. Nash's proposal, in time funded by Lederman's administration, offered a long-term solution to the laboratory's computing problems, but it did not address the serious problems E-516 faced as its members continued to analyze their data with less than adequate computing resources.

Rise of the E-516 Experiment String: The Birth of E-691

To argue for the support of P-691 based on E-516's too limited data remained a tricky business. Meanwhile the group within E-516 that opposed using the recoil detector grew. By the fall of 1982, a few members were exploring a new detector based on the silicon microstrip detector (SMD), a European technology invented between 1978 and 1980 by Robert Klanner. Allowing the simultaneous use of many more electronic detector channels, the SMD could measure particle tracks to an accuracy of 15 microns using fine-grained detectors of silicon mounted to form x, y, and z coordinates. One advantage of the new approach was its offline filter for selecting events with short-lived particles,

113 Thomas Nash, "A Program for Advanced Electronics Projects at Fermilab," 11 May 1982, NP.

such as the D°, a charmed particle produced by the E-516 experiment.[114] But if the data were recorded to tape for later analysis, the new scheme would need even more online and offline computing.

Lederman continued to view E-516's lack of published data with suspicion despite the team's excellent progress in developing the SMD. As it appeared that his continuing pressure on the group to publish its results had had no effect, in March 1983, more than two years after the end of E-516's final run, Lederman resorted to his most powerful incentive. He told E-516 to either publish its data or withdraw its proposal for P-691.[115] This ultimatum made the next three months a tense time. The group's annual progress report, due in May 1983, would be the basis for presenting P-691 at the critical PAC meeting held in June in Aspen, Colorado. Fortunately, after three years of mining their nearly barren data tapes, E-516 now had preliminary results to present. Nash asked for a brief extension of the May deadline to allow the collaboration to present its results in early June to the Friday afternoon "Wine and Cheese Seminar," Fermilab's joint experimental and theoretical physics seminar reserved for presenting the laboratory's most important results.

At this point, the E-516 collaboration had to change the way it worked. During the design, construction, and installation phases, the technical and financial demands of building and commissioning the spectrometer had forced members of the group to focus attention on the spectrometer region for which they had responsibility. Most members gained some knowledge of systems for which they had no direct responsibility, but they typically had only a rudimentary understanding of the entire detector. Now, as the collaboration approached publication, the team had to function as a whole, for reconstructing particle events drew on results from multiple regions of the detector. This process of bringing results together and coming to an agreement on what to publish intensified the interactions between groups, and the boundaries that had earlier modularized the collaboration became less rigid.

The pressure to publish the E-516 data was complicated by the need to proceed rapidly in planning P-691. With funding from Santa Barbara, the preliminary design for P-691 took shape under the leadership of Morrison and his postdoctoral fellow Paul Karchin. Soon another distinguished physicist joined the effort. Morrison had, for some time, been encouraging Michael

114 Paul Karchin, "Silicon Microstrip Detectors for the TPS," 18 Oct 1982, NP.

115 Norman Gelfand to Thomas Nash, 3 Mar 1983, NP.

Witherell of Princeton to join the faculty of the University of California at Santa Barbara and work with E-516. After accepting, Witherell assumed the task of analyzing E-516's data tapes along with a graduate student, Don Summers, with whom Witherell had designed, constructed, and written the reconstruction program for E-516's segmented liquid ionization calorimeter (SLIC).[116]

The first piece of E-516's data that the group submitted for publication would appear in its paper on the decay of the D° particle. Kris Sliwa, a member of the group, presented this at the Friday Wine and Cheese Seminar on June 10, 1983. In his talk, "First Results from the Tagged Photon Spectrometer," Sliwa explained that analysis of the data was in progress and that more data would be announced soon.[117] The P-691 proponents hoped that Sliwa's presentation would ease Lederman's concerns and bring him to grant first stage (conditional) approval for the experiment. Hoping to assure Lederman that E-516's results would soon be published, the group submitted the overheads from Sliwa's talk with their design for P-691, which included hardware improvements in most systems and, despite some objections, E-516's trigger processor. Nash outlined the group's plan to use both the Santa Barbara group's silicon microstrip vertex detectors behind a short beryllium target and, perhaps simultaneously, its liquid hydrogen-filled recoil detector. Nash also revealed that Bintinger's charged track reconstruction program was being used to analyze E-516 data on the new Advanced Computing Program (ACP) resulting from his earlier computing proposal.[118] Although Lederman appreciated this convincing report, he informed Nash two weeks later that P-691's status would remain deferred until he saw concrete evidence that E-516's results would actually appear in a journal. But because Nash insisted that such publication was imminent, Lederman agreed to let him present the plans for P-691 at the November 1983 PAC meeting.[119] Meanwhile, in July, Witherell circulated drafts of his paper

---

116 Don Summers, interview by Mark Bodnarkzuk, 19 Jul 1990, FA.

117 Kris Sliwa, "First Results from the Tagged Photon Spectrometer," overheads prepared for the Wine and Cheese Seminar on 10 Jun 1983, NP.

118 Thomas Nash to Leon Lederman, Taiji Yamanouchi, and Norman Gelfand, 13 Jun 1983; Thomas Nash to Taiji Yamanouchi and Norman Gelfand, 13 Jun 1983, all in NP; Thomas Nash, "Fermilab's Advanced Computer Program," 1984, *Fermilab Annual Report*, FA, 49–55.

119 Leon Lederman to Thomas Nash, 30 Jun 1983, NP.

with Summers on the D°.[120] The plans for publishing E-516's results continued to be coupled with the designing of P-691.

The members, most of them now back at their home institutions, could agree neither on the reconstruction program nor on the design of the new experiment. Multiple drafts were circulated without consensus while the scheduled November presentation to the PAC drew near.[121] By late summer 1983, the group's study of the small sample of charmed particles that had accumulated in the last three months of E-516's final data run had convinced everyone in the collaboration that charm was rarely produced diffractively and that the P-691 configuration should therefore not include the recoil detector. The new plan was supported by Karchin and Morrison's design of a beryllium target/SMD configuration, which would produce the same charmed particle event rate as their opposing group's hydrogen target scheme.[122] The group also agreed on an important change in the collaboration: that Witherell, rather than Nash, should act as the spokesperson for E-691. But they refrained from notifying Fermilab's management of this decision for fear that it might adversely affect the approval of P-691.

Witherell took responsibility for submitting E-516's results, "Study of the Decay $D^{\circ} \rightarrow K^{-} \pi^{+} \pi^{\circ}$" to the journal *Physical Review Letters.* On October 5, 1983, he set October 10 as the deadline for submitting the paper. "It has gone through six complete typings since it was put together in July," he wrote. He said he had tried to reach a consensus on all issues, but "I think it has hit the equilibrium point where for every change I make, I offend as many people as I satisfy." He decreed that from then on, "I will consider only changes that are typos, wrong numbers, or outrageous breaches of grammar inserted when the last changes were made."[123] Then he submitted the paper. Even before the paper was formally accepted, the collaboration informed Lederman that it would be ready to make its November presentation on P-691 to the PAC.[124] Nash was scheduled to present on November 10, 1983; the collaboration reassigned the

120 Michael Witherell to the E-516 Collaboration, 24 Aug 1983, NP.

121 "Inelastic and Elastic Photoproduction of J/Ψ (3097)," draft, 7 Sep 1983, NP.

122 Paul Karchin to the collaboration, handwritten note, 7 Sep 1983, NP.

123 Michael Witherell to E-516 Collaboration, 5 Oct 1983, NP; Donald J. Summers et al., "Study of the Decay $D^{\circ} \rightarrow K^{-} \pi^{+} \pi^{\circ}$ in High-Energy Photoproduction" *Physical Review Letters* 52, no. 6 (1984): 410–13, received 14 Oct 1983.

124 The TPS Collaboration to Taiji Yamanouchi, 15 Oct 1983, NP.

task to Witherell. One week after Witherell's presentation, the three-times-deferred P-691 was granted "Stage I" (conditional) approval. Stage II (full) approval was contingent on Lederman's review of the readiness of the silicon microstrip detector (SMD).[125] Ten days later, Nash sent his formal letter to the E-691 collaboration declaring the "rotation of spokespersons."[126] Nash would continue as a collaboration member, but the job of negotiating the problems that lay ahead would fall to Witherell.

The E-516 String

The original E-516 experiment thus gave rise to an experiment string. It led not only to E-691, but to E-691's follow-up experiments, E-769 and then E-791. The four experiments in the string—E-516, 691, 769, 791—shared a number of structural components. In each, the beam of accelerated protons emerging from the Fermilab accelerator was used to produce either an incident photon or hadron beam. This beam of photons or hadrons would go on to interact in a target and create either photoproduced or hadronically produced charmed particles. E-691 went on to much subsequent success, obtaining 10,000 reconstructed photoproduced charmed particle events and the most precise measurements yet of charmed particle lifetimes. The proposal for E-769, the follow-up to E-691, was submitted to the laboratory in November 1985 and approved in December 1985. The experiment was completed in February 1988, having had a substantial time overlap with its predecessor, E-691. Using a beam of π and K mesons instead of a photon beam to study the charmed particles, E-769 would accumulate 6,000 reconstructed hadronically produced charmed events, the largest such sample in the world at that time. And the proposal for E-791, submitted to the laboratory in November 1987 and approved in June 1988, proceeded to analyze its data in 1992, accumulating 100,000 reconstructed hadronically produced charmed particles. The later experiments in the E-516 string thus greatly transcended the goals of the original experiment in the string, goals which in its time had been ambitious.

Numerous continuities in the E-516 string helped the laboratory economize on precious resources. The transforming of one experiment into another, even before the first had been completed, had made it possible for the collaboration to maintain possession of the TPL. Using the tagged photon beamline

125 Leon Lederman to Thomas Nash, 17 Nov 1983, NP.

126 Thomas Nash to the E-691 collaboration, 28 Nov 1983, NP.

that the E-516 group had built for all four experiments removed the need, and hence the cost, for each of the subsequent experiments to design, construct, commission, and pay for a new beamline. With the TPMS as the basis for all three subsequent experiments in the string, a well-understood detector could be used without expending the funds and time to perform the many tasks involved in commissioning a new spectrometer. Despite the struggle over reconfiguring E-516's trigger, the experiment's detector, the TPMS, showed remarkable structural continuity over the course of the string. And because most of the detector's subsystems remained the same through all four experiments, much of the software for reconstructing particle events could be reused with simple modifications of the code. The most significant difference between the detection apparatus of E-516 and the other experiments was that a silicon microstrip detector (SMD) replaced E-516's recoil detector and trigger processor; a simpler and less restrictive trigger could then be used because more offline computing was available to handle the increased volume of recorded data. (E-691 used somewhat different hardware and software reconstruction packages, enabling it to more accurately study particle tracks and reconstruct the particle decays.) The fact that E-769 and E-791 studied hadronic rather than photoproduced charmed particles had been planned as early as 1976, even before P-516 was approved, for during E-516's conception stage, plans had been made to later convert the original beam to transport pions. And pions were used in E-769's and E-791's studies of hadronically produced charmed particles.[127] By the time P-769 was submitted, it had been ten years since the design for converting the secondary beam to a pion beam had been established. But even using a hadron rather than a photon beam, the configuration for E-769 was so closely based on E-516's that the collaboration included a diagram of the E-516 configuration in P-769 and simply explained the modifications that had been made.[128] By the time of P-791, the design of the TPMS had become so well known to the Fermilab community that the proposal did not even include an image of it. E-691's ability to produce publishable physics results very shortly after

127 The earlier proposal was "Proposal to Study Photoproduction of Final States of Mass Above 2.5 GeV with a Magnetic Spectrometer in the Tagged Photon Lab," submitted 1 Oct 1976, NP.

128 The difference between the two configurations included the addition of two banks of drift chambers to the tracking system, enabling greater redundancy in resolving the particle trajectories, and the upgrading of the online computing capability with a VAX11/780 for data monitoring, so that the PDP11/55, used in E-516 for both data acquisition and data monitoring, could focus entirely on acquiring data from the detector subsystems.

the end of its data run was a direct result of the decision not to modify most of the components of the E-516 spectrometer. Although the TPMS did not become a facility for general use, as had originally been proposed, it did yield a productive and stable long-term program.

The experiment strings had a substantial social effect on the laboratory. With each iteration of the TPMS in the string, the laboratory could see its investment in the TPMS and E-516 bearing more fruit. The investment was seen as increasingly secure as the group developed a track record of success. Experiments in a long-standing string were more easily approved and were assigned higher priority than freestanding experiments. The later experiments in a string usually took less time to yield results, for by using a version of the same detector, the follow-up experiments could avoid spending time on the myriad of tasks surrounding the commissioning of a new spectrometer. In the competitive context of particle physics, the longevity of an experiment was a crucial resource for professors who needed access to high-energy beams to train graduate students. Strings also helped university-based physicists publish results more regularly and thus attract better graduate students. The accumulation of group expertise while doing an experiment became a resource for the follow-up experiments. Thus, in the case of the E-516, 691, 769, 791 string, which spanned two decades from conception to final publication, building the TPMS proved to be a good investment.

Wilson had celebrated risk as part of his frugality initiative, both to cut cost and for aesthetic reasons, but as larger, more complex, and more costly experimental facilities were needed during the Lederman era, from 1979 to 1989, the attitude toward taking risks shifted at Fermilab. Risk became something to avoid rather than to embrace, and strings helped to minimize risk by allowing long-term investments in most of the resources used in the experiments—e.g., protons, equipment, experimental halls, and people with relevant expertise. (Computers tended to be upgraded or replaced regularly and thus did not become a true long-term investment.)

Continuity in the physics goals of the experiments in the string also contributed to a unity of goal and approach of the experiments in the E-516 string. Rather than propose completely new physics measurements and risk rejection by the Fermilab directorate, the follow-up experiments focused on contributing to the continuous program of studying charmed particles. While each data set was used separately, the collaborations accumulated increasingly larger samples described by increasingly precise statistics, enabling more fine-grained measurements of charmed particles.[129] The experiment strings thus defined stable research traditions within the laboratory; in the case of the E-516 string, a tradition emerged for studying the production and behavior of charmed particles.

The constitution of the groups also showed continuities, but for the E-516 string this was a less striking feature than in other Fermilab strings. Appel worked on all four of the experiments in the E-516 string, but Nash, Morrison, and Nauenberg left the string at some point. A few, like Summers, left and returned to the string, while others, like Witherell, joined at a later stage.

Strings as a Limiting Process in Megascience

The continued growth of high-energy physics experiments in the 1970s and 1980s, in a context of limited resources, changed the way physicists worked, especially when their work was constrained by a philosophy of frugal use of its resources. Competitions arose between subgroups within the same experiment collaboration, as well as between different experiment groups. At stake were a variety of resources, from apparatus and working space to particle beam or computer support. In the case of E-516, the decision to conduct a high-statistics experiment in a limited computing economy almost caused E-516 to fail. The most dramatic outcome of these competitions was the formation of experiment strings as a way to retain or extend precious resources. By the time Lederman stepped down as Fermilab's director in 1989, experiment strings had become the relevant unit for research in Fermilab's experiment program.

The resources for experiments were negotiated in diverse economies within the laboratory. One economy was concerned with the protons that the laboratory made available to experimenters. The proton economy was one in which the demand always exceeded the supply. Once a team gained access to the proton beam at the end of a beam spigot, it set up a defensive posture in its effort to retain its access to the proton beam. Other resource economies at the laboratory centered on physics expertise, experiment real estate (both laboratory space and apparatus), computer support, and journal space. As a commodity, physics expertise tended to appreciate in value as the complexity of the physics and the experiments increased. A proper distribution of expertise

129 The first hadron experiment, E-769, actually demonstrated less charm than E-691, but it proved techniques that made hadron experiments more tractable.

was needed to design, fabricate, install, and operate both the apparatus and the computing systems.[130] And once a person had invested the time and effort to become trained in a specific line of physics, it made sense to have this training serve him or her in related experiments.

Experimental real estate, including the experiment hall and the detectors, became an increasingly important subject of the wars within the laboratory, because the time needed to assemble and commission an experiment grew longer as detectors grew larger and more complex. Having invested the time, funds, and effort to develop and then erect a complex detector in an experiment hall, a group could not afford to give up this real estate. Thus, when a group was awarded beam time it would move into an experiment hall with the explicit goal of performing an experiment and the implicit goal of not moving out for a long time. The size and complexity of the detectors began to shape the direction of the physics as well as the social structure of the field as the members of experiment groups matched their training and interests to the needs of the detector's subsystems. Physicists became typed according to the kinds of detectors with which they worked. Experiment proposals were judged, at least in part, by whether they could attract physicists, or students, having the right expertise for a particular experiment. This process even affected the rate of production of PhD candidates having the appropriate backgrounds for working on specialized aspects of experiments.

Computer facilities were a crucial resource, as our study illustrates. The more secure "write-it-all-to-tape" approach produced much more data but required immense offline computing resources, which were simply not available in Wilson's frugally run laboratory. As computer resources were frequently updated, they had a less conservative influence than the other resources we have mentioned. Finally, the economy of journal space for published results played into the picture, because the laboratory saw publications as a return on its investment in the research. The laboratory director's power to approve or disapprove an experiment based on publication of previous work was a powerful management tool, and its influence was magnified by the fact that physicists recognized publications as the basis for being authorized for follow-up experiments and also for obtaining jobs, grants, raises, and promotions.

130 A study conducted by the High Energy Physics Advisory Panel (HEPAP) examining the United States research program in high-energy physics for the 1990s included a demographic study of "manpower considerations," "The U.S. High Energy Physics Research Program for the 1990s," *HEPAP Subpanel Report* DOE/ER-0453P, Washington, DC, Apr 1990, 68ff, FA.

This study of how E-516 gave birth to an experiment string suggests that strings are the product of a tension felt by any large and expensive experiment conducted in a context where the resources are limited. This tension results from the opposition between a research effort's tendency to grow larger when adequately funded and the constraining influences that occur when resources are limited. With the possible exception of computing, all the resource economies of the laboratory contributed to the continuation of experiments beyond the duration they might have had in more bountiful times. While the particle physicists' desire to explore the frontiers of their subject propelled them to investigate new scientific territories, limitations on resources constrained their studies to a narrowing region of the frontier. The emergence of strings in the programs of experiment at large high-energy physics laboratories can thus be viewed as a limiting process in the evolution of large-scale research.[131] The experiment strings helped curb the unlimited growth of science that had concerned Derek J. De Solla Price, Alvin M. Weinberg, and others who wrote in the early 1960s about the worrisome implications of big science.

Acknowledgements

For helpful conversations, references, criticism, and access to documents, we are grateful to a large number of Fermilab physicists, especially Thomas Nash and Jeff Appel. For constructive criticisms we thank the following members of the continuing Laboratory History Conference: Robert Crease, John Krige, Stewart Leslie, Michael Riordan, Robert Seidel, and Catherine Westfall. For countless inputs to this paper, ranging from archival research to historical analysis and editing, we are deeply indebted to Adrienne Kolb, the Fermilab archivist. We also thank Fermilab for opening its doors to this research and encouraging its execution in many ways.

131 Similar continuities to those noted for the E-516 string at Fermilab can be seen in other strings at Fermilab, e.g., (E-82, 226, 425, 486, 584, 617, 731, 773, 799), (E-8, 361, 415, 440, 495, 505, 555, 619, 620, 756, 800), (E-21A, 262, 320, 356, 482, 616, 652, 701, 744, 770), (E-1A, 310), (E-537, 705, 771), (E-70, 288, 494, 605, 608, 772, 789), (E-36a, 186, 317, 381, 289), and (E-87, 358, 400, 401, 687, 831). At NAL's 15-foot bubble chamber, the major continuity seems to have been the chamber itself, used in (E-28A, 31A, 45A, 53A, 155, 172, 180, 202, 234, 341, 380, 388, 390, 502, 545, 546, 564, 632). Experiment strings emerged at other high-energy physics laboratories in the same time period as the E-516 string, e.g., the photoproduction experiments at CERN (NA1-1980 run, NA1-1983 run, NA14, NA14/2), as well as (WA4, WA57, WA58, and WA69).

Fermilab

Fermi National Accelerator Laboratory
P.O. Box 500 • Batavia, Illinois • 60510

Directors Office

Dear colleague, May 5

An all too common failing of large institutions is to fall into the bureaucratic morass — complicated procedures, red-tape, and all that. That's terrible.

Lets try hard to keep the good old can-do informal spirit of Fermilab alive! I ask each of you to be intolerant of creeping bureaucracy.

Bob Wilson

Robert Wilson resigned as Fermilab's first director on February 9, 1978, because of the political and bureaucratic difficulties of securing adequate funding to raise the proton beam energy in the accelerator from 400 GeV to 800 GeV with the proposed Energy Doubler project. Wilson agreed to stay on as Fermilab's director until a replacement was found, but he became increasingly frustrated by the ongoing difficulty of providing creative leadership to Fermilab and its staff. On May 5, 1978, Wilson's frustration erupted and he wrote and distributed copies of the handwritten letter shown above to all Fermilab employees (See Lillian Hoddeson, Adrienne Kolb, and Catherine Westfall, *Fermilab: Physics, the Frontier, and Megascience* (Chicago, IL: University of Chicago Press, 2008), 214-217.)

3

The Advent of Management and Removing the Uncertainties in High-Energy Physics Experiments[132]

Abstract

Physicists in high-energy physics (HEP) have been forced to organize large scientific projects without a well-defined organizational or sociological model to guide them. In the absence of such models, what structures do experimentalists use to develop social structures in HEP? In this article, I claim that physicists organize around what they know best, the physics problems they study and the detectors and devices they study them with. After describing the advent of "management" in HEP, I use a case study of four Fermilab experiments to propose a physics and detector driven model of social structure for experiments. In addition, I show how this model can be extended to describe "strings" of experiments, where continuities of physics interests, spectrometer design, and a core group of physicists become a definable sociological unit that can exist for more than 15 years. A dominant theme that emerges from my analysis is the conscious attempt on the part of experimenters to remove the uncertainties that are part of the practice of HEP at large laboratories like Fermilab.[133]

---

132 This article was originally published by Fermi National Accelerator Laboratory as Fermilab as, Mark Bodnarczuk, *The Social Structure of "Experimental" Strings at Fermilab; A Physics and Detector Driven Model*, TM-1706, December 12, 1990.

133 Some of the ideas presented in this paper were presented in the form of an invited talk given at Lawrence Berkeley Laboratory on July 10, 1990. The vast majority of the material is part of my ongoing PhD dissertation research at the University of Chicago under the direction of Daniel Garber and William Wimsatt of the Philosophy Department and Bruce Winstein of the Enrico Fermi Institute. In addition, continuing discussions with Lillian Hoddeson, Catherine Westfall, Adrienne Kolb, and

The Advent of "Management" in HEP

In the days of cosmic ray experiments, the world was a laboratory in which experimenters set up their detectors in the path of incident particle beams from stellar activities in deep space. But the particle flux from cosmic rays could not be predicted and long detector exposures on high mountain tops (or on ships) made it difficult to accumulate large samples of particle interactions.[134] Early "machine builders" designed and constructed accelerators for the purpose of freeing experimentalists from the uncertainties inherent in performing cosmic ray studies.[135] The ability to produce particle beams of a defined energy and intensity at will enabled them to be free of the intermittent cosmic ray particle beams and accumulate much larger data samples.

With the advent of particle accelerators, experimentalists began setting up their detectors at particle accelerator laboratories in the path of incident particle beams produced by accelerators. In the 1950s, accelerators had circumferences of order 63-126 meters and could be housed in large buildings.[136] The discovery of strong focusing by Courant, Livingston, Hartland, and Snyder, enabled the size of accelerators in the 1960s to grow larger, with circumferences of order 628 meters that were housed underground in tunnels.[137] Machines in the 1970s (like Fermilab's Main Ring) were about 6 kilometers in circumference, reaching

---

Kyoung Paik under the auspices of the Fermilab History Collaboration have been very helpful in formulating some of my ideas.

134 See H. Victor Neher, "Some Reminiscences of the Early Days of Cosmic Rays" in Laurie Brown and Lillian Hoddeson (eds.), *The Birth of Particle Physics* (New York: Cambridge University Press. 1986) 120 ff, and Carl D. Anderson and Herbert L. Anderson, "Unraveling the Particle Content of Cosmic Rays," in Laurie Brown and Lillian Hoddeson (eds.), *The Birth of Particle Physics* (New York: Cambridge University Press, 1986) 131 ff.

135 Heilbron and Seidel claim that the main motivation of early "cyclotroneers" like Lawrence was the "macho character" of building bigger and bigger machines and they were only secondarily concerned about the experiments done with those machines. See J. L. Heilbron and Robert W. Seidel, *Lawrence and His Laboratory: A History of the Lawrence Berkeley Laboratory*. vol. 1 (Berkeley, CA: University of California Press, 1989), 466.

136 The Cosmotron at BNL had a beam energy of 3 GeV, while the Bevatron at the Lawrence Radiation Laboratory could reach energies of 6.2 GeV.

137 The Alternate Gradient Synchrotron at Brookhaven National Laboratory could reach energies of about 33 GeV, the CERN Proton Synchrotron about 28 GeV, and the "weak" focusing Zero Gradient Synchrotron at Argonne National Laboratory could reach energies of about 13 GeV. See Laurie M. Brown, Max Dresden, and Lillian Hoddeson, "Pions to Quarks: Particle Physics in the 1950s" in Laurie M. Brown, Max Dresden, and Lillian Hoddeson (eds.) *Pions to Quarks; Particle Physics in the 1950s, Based on a Fermilab Symposium* (New York: Cambridge University Press, 1989), 13 ff.

energies between 200-400 GeV, and with the advent of the superconducting Tevatron, reached energies of 1.8 TeV in the center of mass in the early to mid-1980s.[138] In the mid- to late 1980s, accelerators like the Large Electron Positron Collider (LEP) at CERN grew to about 27 kilometers in circumference, and the SSC (designed for operation in the next millennium), has a circumference of about 90 kilometers and will reach energies of 40,000 GeV in the center of mass.[139] Because the required circumference of the accelerator is approximately inversely proportional to the B-field strength of the magnets used in the machine, the rising masses of the particles to be studied and the falling cross sections for particle production demanded the size and complexity of the machines built also increase. But rising particle masses and falling cross sections for particle production also meant a substantial rise in the cost of building these larger and larger machines. Thus, the resources for doing HEP research were localized in a few institutional laboratories, creating a monopoly that came to be funded largely by the United States government.[140] The claim that the size, complexity, and cost of these accelerators was powerfully driven by the need to penetrate into new physics domains is clear. But what is much less clear is that the size, complexity, and cost of these machines also demanded the development of more complex social structures to manage the monopoly that laboratories had on the resources for doing HEP.

The ability to produce particle beams of a defined energy and intensity at will enabled experimentalists to be free of the uncertainties of naturally produced cosmic rays, but this thrust them into a new type of sociological uncertainty produced by the monopoly that accelerator laboratories had on particle beams. For scientists in HEP, such social factors are sometimes more

---

138 The Fermilab Main Ring was originally designed as a 200 GeV machine that later ran at 400 GeV. See Catherine Westfall's, *The First "Truly National laboratory": The Birth of Fermilab*, 2 vols, unpublished dissertation submitted to Michigan State University, Department of English, 1988, vol 1, 160 ff.

139 My primary focus here is the size of accelerators, not the particle type, thus I have included LEP in what is otherwise a description of hadron colliders.

140 J. L. Heilbron and Robert Seidel discuss how the development of accelerators was inextricably tied to government funded weapons research. See J. L. Heilbron, "An Historians Interest in Particle Physics," in Laurie M. Brown, Max Dresden, and Lillian Hoddeson (eds.) *Pions to Quarks; Particle Physics in the 1950s, Based on a Fermilab Symposium* (New York: Cambridge University Press, 1989), 50 ff, and Robert Seidel, "Postwar Political Economy of High-Energy Physics," in Laurie M. Brown, Max Dresden, and Lillian Hoddeson (eds.) *Pions lo Quarks; Particle Physics in the 1950s, Based on a Fermilab Symposium* (New York: Cambridge University Press, 1989), 497 ff.

difficult to navigate than the capriciousness of incident cosmic ray particles. The rise of this institutional monopoly on resources gave birth to the large scale "management" of HEP. The monopoly on resources subsequently produced three new and powerful factors: proton economics, experimental real estate, and a redefinition of what it meant to be a particle physicist in HEP. The first factor, proton economics, was vitally constrained by the physical parameters of the accelerator's ability to produce a given number of protons and the cross section for particle production. The sociological and management parameters emerged when considering how to best utilize these resources, namely the choice of which types of particle beams to produce and how to distribute them among the experimental community at a laboratory like Fermilab.

The parameters of proton economics involved numerous constraints, the first of which was the cross section for secondary and tertiary beam production. Given the total number of protons from the accelerator, the laboratory had to decide what type of secondary beams to provide for experimenters in light of the experimental proposals that were submitted. For example, because the cross section for neutrino production is 10^{-36} cm^2 and the pion cross section is 10^{-27} cm^2 the decision to approve experiments that used incident beams of neutrinos was already a major decision that affected proton economics.[141] It was clear that a neutrino beam was much more costly than a pion beam in terms of the number of protons needed, especially if the intensity of the secondary beam was of large order.

The second constraint of proton economics was the cross section for particle production in the experimental target. The decision to propose an experiment had to be proceeded by a Monte Carlo study of how many potential events could be detected in the experimental spectrometer with a given secondary beam intensity. By knowing the secondary beam intensity, the repetition rate, the physical properties of the target (luminosity), and the cross section for the interaction to be studied, the experimenters could calculate a reaction rate for a particular type of particle event. Once the reaction rate was calculated, the experiment determined the number of events needed to provide adequate statistics in their data sample, given a specific number of beam hours. When proposing an experiment, the collaboration began by calculating the desired

141 The pion cross section is roughly constant for energies above 2 GeV at about 40 millibarns. The neutrino cross section is not constant but is linearly proportional to the energy. For Fermilab, a reasonable neutrino energy to use is 100 GeV, which would give a cross section of about 0.7 picobarns.

reaction rate and working back through the process to determine the number of incident protons needed on the primary production target and they included that number as part of the experimental proposal.[142]

The rise of the institutional monopoly on the resources to do HEP led to a second powerful sociological factor—experimental real estate—something related to proton economics but less defined. Given the size and complexity of experimental detectors and the long lead times needed to assemble, operate, and analyze the data, collaborations of physicists who were given beam time moved into the experimental hall at the end of one of the particle beam spigots with the explicit goal of not moving out. The experimental proposals were based upon the collaboration using general purpose equipment provided by the laboratory. Only portions of the spectrometer were built using the resources of the collaborating universities themselves. In some instances, the laboratory invested large amounts of resources into more general-purpose facilities that, with minor modifications, could be used to do a series of different experiments.[143]

The phenomenon of experimental real estate manifested itself in a number of ways, all of which were aimed at not moving out of the enclosure once a collaboration secured that valuable piece of property. Incumbent experiments could try to convince the laboratory of the importance of the potential physics with the goal of obtaining more laboratory resources for a follow-up experiment. The resources were not just beam time, but often involved the resources needed to upgrade the spectrometer. If they succeeded, they could leave the major components of their spectrometer in place and gain additional access to the beam. If the above strategy failed and other experimental groups threatened to displace them by submitting a proposal that used that beamline and enclosure, the next best strategy was for the incumbent experiment to try to convince the laboratory (and the intruding experimenters) to use portions of

142 For example, a rough summation of all the intensity requirements for experiments proposing run during the 1993 fixed-target run at Fermilab is about $5X10^{13}$ protons. The accelerator is currently capable of producing about $1.5X10^{13}$ protons. The parameters of proton economics will vitally constrain the total number of proposed experiments will actually be approved to run.

143 An early example that I will focus on as a case study, is the Tagged Photon Magnetic Spectrometer at the Tagged Photon Lab. But with Leon Lederman's decision to propose Tevatron II (an upgrade of the fixed-target experimental areas to transport 800 GeV beams which included the construction of a number of large experimental facilities), Fermilab began to place more emphasis on building general purpose spectrometers that could be used to exploit an entire program of physics experiments.

their spectrometer as part of the invading experiment's spectrometer configuration (something which the competing group argued against). Doing this enabled the incumbent to partially "hold his place in line," and increased the probability that the laboratory would allow them to do subsequent experiments. In generating their arguments against giving up the real estate, the incumbent experiment would use the data from their previous run and the promise of new results to show that their experiment was "more important" to the progress of HEP than the proposed results of the experiment seeking to displace them.

The rise of the institutional monopoly on the resources to do HEP produced a third sociological factor, which involved a redefinition of what it meant to be a physicist in the discipline of HEP. This manifested itself as an increasingly complex division of labor where the major responsibility of some physicists was to provide the laboratory support, which allowed other physicists to perform experiments. These "service" jobs involved activities like the design, construction, and operation of the accelerators that produced the incident beams, the external beamlines used to transport those beams to the experimental apparatus, and the computing support needed to analyze the data from experiments. Thus, the purpose of the physicists who constituted this new support structure was to make the resources of HEP available, but to whom?

This took two directions depending on the operating philosophy of the laboratory. Laboratories like Brookhaven National Laboratory (BNL) and the Lawrence Berkeley Laboratory (LBL) had strong "in-house" groups that maintained these systems and exerted control over the resources, making it difficult for "outside" users to gain access to particle beams. Within the physical parameters of proton economics, laboratory management decided to distribute the beams to users who were affiliated with the laboratory institution. Thus, there was little or no distinction between the social structures that made the resources of HEP available, and the social structures of the experimenters who used those resources. At Fermilab, there was a pronounced division between physicists who provided the resources and the outside users who wanted access to the beams. The redefinition of what it meant to be a physicist in HEP was seen most profoundly at Fermilab because the in-house/outside-user distinction was built so deeply into Robert Wilson's image of laboratory life.[144] With the in-house style of manage-

144 Wilson was committed to what Lederman called a "TNL," a truly national laboratory, the acronym of which was a phonetic play on BNL where in-house groups dominated the resources for doing HEP. As Wilson remarked, ". . . NAL leaders intended to make outside users "the whole focus" of the laboratory, in keeping with the spirit of Lederman's "Truly National Laboratory." See C.

ment, affiliation with the institution was the best strategy for securing resources. With the Fermilab style of management, it was more important to convince laboratory management that the physics was worth doing and that the collaboration had the expertise and manpower to actually perform the experiment. While the two management approaches are markedly different, they are clear examples of the increasingly complex social structure and management needs of HEP. Unlike the marked absence of such factors in cosmic ray experiments, navigating the social negotiations involved in proton economics and experimental real estate became at least as important as anything "experimental" in the actual process of beginning, operating, and ending an experiment.

A Physics and Detector Driven Sociology: A Case Study

The emergence of the higher-level sociological structures needed to manage the resources of HEP at the institutional level, also created less understood lower-level issues for the social structure of the experiments that utilized those scarce resources. These social factors became as important as anything "experimental" when performing an experiment. This case study shows that in the absence of a formal sociological model for structuring collaborations, physicists organized themselves around what they knew best, the physics problems they studied and the spectrometers they studied them with. More specifically, a definable method for analyzing the social structure of collaborations emerges from a description of their physics goals and a diagram of the spectrometer—it's a physics and detector driven sociology for HEP experiments. In what follows, I briefly discuss the history of the development of detectors, then use the example of experiment E-516 (Photoproduction of Charm) and other experiments which ran at Fermilab.

The historical development of detectors used for HEP moved along two definable but orthogonal axes, both of which were the direct result of the nature of the physics problems being studied. Along the first axis, detectors increased in size, but not necessarily in architectural complexity. An example highlighted by Peter Galison was the development of the table-top sized spark chamber by Cronin and Renninger in 1960, and its expansive extrapolation by Melvin Swartz and Leon Lederman into a device that was large enough to be

Westfall, "Fermilab: Founding the First U.S. 'Truly National Laboratory'," in F.A.J.L. James (ed), *The Development of the Laboratory: Essays on the Place of Experiment in Industrial Civilization* (London: Macmillan Press, 1989), 184-217, 331 ff.

sandwiched between surplus naval cruiser deck plates weighing between two and three thousand tons.[145] The extrapolation from spark chambers that were table sized to the mammoth detectors used by Lederman in the two-neutrino experiment performed at BNL or the Fermilab based E-1A experiment performed by Rubbia, Cline, and Mann was necessitated by the fact that they were both performing studies on neutrino interactions.

The cross section for neutrino production necessitated the larger target mass along with the increased requirements for shielding the fiducial volume of the detector. Consequently, the design of such experiments was inextricably tied to the physical properties of the neutrino cross section and the cross sections of the hadrons and leptons that composed the background for weak neutral current events.[146] But because detecting neutral currents only involved detecting the presence or absence of a muon track protruding from the cluster of hadrons produced in the incident neutrino collision, the E-1A detector, for all its size, was relatively uncomplex having only 46 spark chambers attached to 46 channels of electronics. Growth in the size of the detectors also meant an increase in the cost of experiments like E-1A. This made collaborations even more dependent on the laboratory for the resources needed to construct and operate these mammoth structures.

In the orthogonal axis of complexity, many varieties of detector types were combined to form increasingly complex spectrometers composed of a variety of detector subsystems. These complex spectrometers enabled experimentalists to record more complex physical interactions and measure increased numbers of physical parameters simultaneously. The Tagged Photon Magnetic Spectrometer (TPMS) used for experiment E-516 at Fermilab was an early example of a spectrometer which grew substantially along both axes, being large in size as well as complexity.[147] The goal of the experiment was to study photoproduction of final states with masses above 2.5 GeV. The spectrometer was an array of individual detector subsystems designed to function as a unit,

---

145 Peter Galison, "Bubbles, Sparks, and the Postwar Laboratory," in Laurie M. Brown, Max Dresden, and Lillian Hoddeson (eds.) *Pions to Quarks; Particle Physics in the 1950s, Based on a Fermilab Symposium* (New York: Cambridge University Press, 1989), 235-237.

146 A careful account of the E-1A experiment has been written by Peter Galison, *How Experiments End* (Chicago: The University of Chicago Press, 1987).

147 See the Tagged Photon Magnetic Spectrometer; Facility Design Report, May 9, 1977. Other examples at Fermilab were the Multi Particle Spectrometer used for experiment E-557 and the spectrometer used for experiment E-400 in the Broad Band Photon Beam Line.

summed at a higher level by a specially designed computing architecture.[148] Unlike the binary signature in the E-1A detector, fully constructing the multiple decay products from charmed particle events required the TPMS to be a significantly more complex apparatus.

As shown in Figure 1, the spectrometer was divided into six detector regions: a) the experimental target/recoil detector, b) the tracking system, c) the Cerenkov counter, d) the segmented liquid scintillation shower counter (SLIC), e) the hadromagnetic calorimeter, and f) the muon system.[149] The complexity of the particle interactions was mirrored in the complexity of the spectrometer which had 7,000 interaction points distributed over the six detector regions, each of which was connected to a channel of electronics.[150]

What this points to is a strong causal relationship between the complexity of the physics interactions and the complexity of the detectors needed to detect those interactions. It also points to the fact that the cost of building such spectrometers substantially increased by virtue of its size and by virtue of its complexity, because more channels meant more money.[151] When spectrometers became as large and complex as the TPMS, the institutions that composed the collaboration not only became more dependent upon the laboratory for financial resources, they also became more interdependent upon one another for bearing their share of the cost of the experiment. The comparison of the E-1A experiment and experiment E-516 points to a physics driven detector design. For those who are familiar with the scientific practice of HEP, this conclusion ought not be surprising.

---

148 Experiment E-516 used a ECL trigger processor that was specially designed by Tom Nash and Steve Bracker. The ECL trigger processor was described in an article written in the May 1983, issue of *Physics Today*.

149 See the diagram of the E-516 configuration of the TPMS in the *Tagged Photon Magnetic Spectrometer; Facility Design Report*, May 9, 1977.

150 Physicists normally refer to an "interaction" point as a point in the experimental target. What I am referring to are points within the detector subsystems where the decay products from particles produced in the experimental target interact, for example, the cell of a drift chamber, or the wire of a proportional wire chamber (PWC).

151 A rough rule of thumb for estimating the cost of a spectrometer is to quantify cost in terms of dollars per channel. In a private communication with Ken Stanfield who designed the drift chambers for the E-516 tracking system, Stanfield claimed that one of the major factors in determining the number of wires/channels that would constitute the tracking system was dividing the dollar amount of the E-516 budget for tracking by the cost per channel.

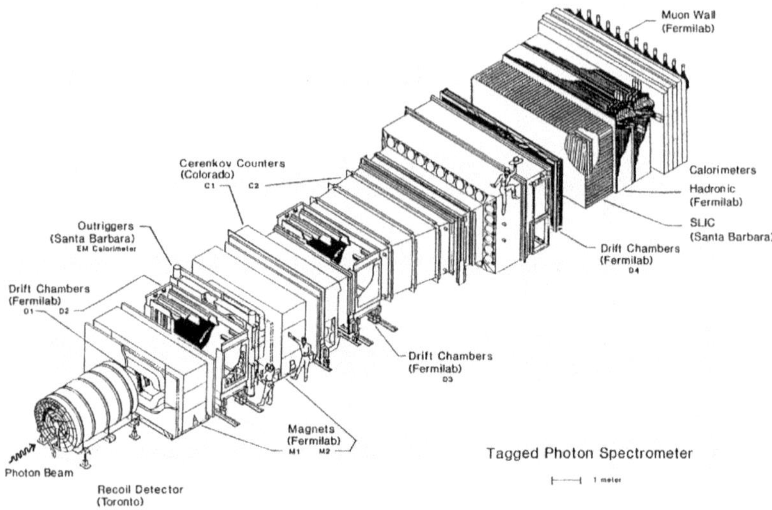

Figure 1: The Tagged Photon Magnetic Spectrometer.

What may be surprising is that the physics driven explanation of detector design does not stop with the detector but extends to the sociological structure of the collaboration itself. In the absence of any systematic sociological model for organizing the E-516 collaboration, the physical structure of the TPMS became the model that shaped and defined the social structure of the collaboration. Each of the six spectrometer regions was the domain of an institutional group from the collaboration and represented their hardware contribution to the overall experiment. The recoil detector was built by the University of Toronto, the SLIC was built by the University of California at Santa Barbara, the Cerenkov counters were built by the University of Colorado at Boulder, and the tracking system, hadronic calorimeter, and muon system were built by Fermilab scientists.[152]

In other words, the basic social structure of experiment E-516 emerged from the institutions identified in the drawing of the TPMS shown in Figure 1, each of which had an institutional representative, and this formed the basic social structure of the E-516 collaboration. So the complexity of the physics

152 The ECL trigger processor and the online computing system were collaborative efforts between Fermilab and the University of Toronto contingents of the collaboration.

interactions determined the physical structure and complexity of the TPMS, and the physicists who carried out the experiment organized themselves around that structure by institution and area of expertise. As spectrometers like the TPMS were carved up into smaller, more manageable pieces, the phenomenon of "technical specialization" emerged in a more powerful way. The physics and detector driven model for E-516 can be used to describe all phases of the experiment, from design, construction, and installation, to running the experiment, followed by data reconstruction and analysis, the preliminary presentation of data by collaboration members, and the final publication of scientific data in peer reviewed journals.

In the design, construction, and installation phases, the physical and financial demands of getting the spectrometer built and operating forced collaboration members to focus their attention on their assigned region of the spectrometer. The detector subsystems were typically constructed at the home institutions then shipped to Fermilab for installation, so there was less social interaction between collaboration members in the earlier phases of the experiment.[153] The collaboration's level of social interaction significantly increased when they came together at Fermilab during the installation phase. Once the spectrometer was installed, the level of social interaction within the collaboration continued to increase as test-beams were used to calibrate the six regions of the spectrometer, and then subsequently in the actual data taking phase of the experiment, which involved 24x7 operations with a rotating shift schedule.

During data taking phase, the social structure that emerged from the detector subsystems defined how technical problems and challenges were addressed by the collaboration. For example, problems in the operation of the SLIC were the responsibility of the Santa Barbara contingent who designed and constructed that detector subsystem, whereas problems with the Cerenkov counters were the purview of the University of Colorado. To be sure, various collaboration members gained a limited knowledge of all the detector subsystems, but the major repositories of expertise for running, calibrating, and fixing the spectrometer's subsystems were distributed along the lines shown in the TPMS diagram.[154]

---

153 There were of course design meetings and reviews performed both within the collaboration and in the context of the laboratory, but the actual detector components were often constructed at the home institution and subsequently shipped to Fermilab.

154 The reality of "technical specialization" where institutions had responsibility for specific detector

The data reconstruction phase of the experiment was spread over a two-year period, which spanned three distinct data runs.[155] During this phase, the well-defined structure of technical expertise based on detector subsystems was relaxed, as various members who had responsibility for one specific detector subsystem developed data reconstruction packages for other regions of the TPMS. In fact, there were multiple data reconstruction packages written for each of the TPMS detector subsystems.[156] As such, the firm boundaries defined by detector subsystem ownership were more easily crossed, as collaboration members sought to build expertise in detector subsystems that were the responsibility of another part of the collaboration. While the foundation of the collaboration's social structure remained anchored in the spectrometer design throughout all experimental phases, the case study of E-516 shows that they became less rigid in the later phases when tasks involved the development of software rather than hardware.[157]

In the final data analysis and publication phase, the collaboration faced the difficult social issues of negotiating the final production of data for publication in a journal. But even then, the detector was the organizing factor around which the collaboration addressed these sociological issues. For example, the primary expertise for data issues involving the SLIC were handled by the Santa Barbara group that designed and built that part of the spectrometer, while the expertise for the Cerenkov counter was in the domain of the Colorado group.[158] Much like the six detector regions of the TPMS had to work together as a unit to detect and then reconstruct the decay products of charmed particles, so also the collaboration members who were distributed along the boundaries

subsystems meant that collaboration members did not have an in-depth knowledge of all parts of large complex spectrometers like the TPMS. Over time, this became increasingly true of Fermilab's large collider experiments like CDF and D0.

155 The first shakedown and calibration run of E-516 was in the summer of 1979, with the first data run beginning in the late fall of 1979. The final data run began in March of 1980 and ended at the accelerator shutdown in June, 1980.

156 For instance, there were three separate reconstruction packages generated for the tracking system, and two packages generated for the SLIC. This was a common scientific practice for HEP experiments in this time frame.

157 Part of this is due to the inability of the collaboration to secure the resources necessary to construct multiple detector subsystems. The resources needed to produce software are much easier to obtain because they only involve the time and expertise of the students and physicists on the experiment which are directly under their control and have little or no financial impact.

158 Analysis of the meeting minutes of the collaboration also support this view.

of the TPMS's detector subsystems were forced to work together as a unit to negotiate the best way to present the data for final publication.

In addition to the social structures that were shaped by the TPMS itself, there was a higher-order layer of social structure that is important to describe. The laboratory requires experiments to appoint a spokesperson who is the liaison between Fermilab and the collaboration in all matters to do with the experiment. The "management" style of the spokesperson is also an important element in social negotiation that occurs within the collaboration and between the experiment and the Directorate. Almost without exception, it is the spokesperson and the laboratory Directorate that are the major players in what might be characterized as the Director/experiment relationship. Independent of the spokesperson's management style (autocratic, democratic, etc.), the underlying social taxonomy of the spectrometer persists as the foundation of the collaboration.

For example, shortly after the laboratory's approval of the follow-up experiment to E-516 (E-691) there was a rotation of spokespersons for the collaboration.[159] But the collaboration's basic social structure continued to mirror the physical design of the TPMS. In addition, the modifications needed to create the new E-691 configuration of the TPMS were assigned to an existing part of the collaboration who had the expertise needed to design, construct and operate that part of the spectrometer.[160] So in the follow-up experiment, the configuration of the spectrometer continued to define the social structure of the collaboration and over time this led to "strings" of experiments, with even more complex and long-lived social structures.[161]

---

159 Tom Nash, spokesperson for E-516, rotated the position of spokesperson with another member of the E-516 collaboration, Mike Witherell, to devote his time to the development of the new low-cost high-power parallel processing computers which he proposed to the laboratory a year earlier. See "A Program for Advanced Electronics Projects at Fermilab" by Tom Nash, May 11, 1982. This proposal eventually developed into the Advanced Computing Project (ACP) under the leadership of Nash.

160 As I will describe in more detail in the next section, the addition of the Silicon Microstrip Detector was the responsibility of the Santa Barbara group.

161 As I will describe in detail in the next section, the only major structural addition to the TPMS was the Silicon Microstrip Detector which replaced the target and recoil region of the E-516 configuration.

An Experiment String Model

Experiments E-516 and E-691 met at an interface where the first experiment transformed into a second experiment.[162] Most physicists in HEP understand this process intuitively and can identify the components of the transformation process. But these intuitions have yet to be recorded and analyzed in a systematic way by physicists, historians, or sociologists of HEP, especially the experimental continuities that constitute such "strings."[163] In what follows, I describe the continuities between multiple experiments like E-516, E-691 and their progeny using the parameters of the physics and detector driven model, including the ability of the core group of physicists to work together as a unit in a continuing program of physics studies that yielded successful results.

The configuration of the TPMS for E-516 and E-691 was almost identical, but there were a few important differences between them.[164] The most important difference was the use of the Silicon Microstrip Detector (SMD) instead of the recoil detector along with more loosely defined trigger assumptions.[165] The SMD allowed E-691 to reconstruct the decay vertices of charmed particles in the upstream region of the spectrometer and project the tracks of particle decays to the forward regions of the spectrometer. Experiment E-516 triggered the TPMS with the upstream recoil detector and projected the particle tracks back toward a vertex that could not be as carefully defined by the recoil detector. In addition, E-691 made a significant addition to the online

162 Experiment E-691 was formally proposed to the laboratory on February 4, 1981, during the reconstruction and analysis phases of E-516.

163 While Peter Galison provides a veridical account of the role that experiment plays in the development of scientific knowledge in HEP, he is more interested in the issue of "realism" and sociological considerations are beyond the scope of his work. See Peter Galison, *How Experiments End* (Chicago: The University of Chicago Press, 1987).

164 Unlike the 400 GeV primary proton beam that serviced E-516, experiment E-691 ran with an incident proton beam of 800 GeV from the newly completed Tevatron. This allowed the photon energy that was incident on the experimental target to be increased from about 160 GeV (for the E-516 configuration) to about 300 GeV (for E-691).

165 Whereas E-516 used the recoil detector in combination with tightly defined trigger assumptions to look for the charm signature of a recoil proton, E-691 used a more loosely defined high Pt trigger in combination with the Silicon Microstrip Detector. It is interesting to note that the spectrometer configuration was so similar to the one used for E-516, that the collaboration simply attached a copy of the original Tagged Photon Magnetic Spectrometer: Facility Design Report produced for E-516 in 1977 to their proposal for E-691 (Proposal to Do Photon Physics with the Tevatron at the Tagged Photon Spectrometer). This included the same spectrometer diagram used for E-516 and the differences between the two configurations were simply described in the text.

computing capabilities, using a VAX 11/780 for data monitoring, and the PDP11/55 solely for data acquisition.[166] The SMD and the substantial increase in online computing capabilities were the only major changes between the two experimental configurations.[167]

In regard to the continuity of physics goals, the original proposal of P-516 already included an ambitious physics program for the spectrometer that transcended the studies done by E-516. The plans for using the TPMS in the E-691 Tevatron configuration were already explicitly stated as early as 1976.[168] The proposal states, "In fact, although this is by no means a proposal for the Energy Doubler, this spectrometer will be ideal—and unique—for studying photoproduction when the Doubler comes into operation. It is probably the only existing P East facility that will be able to operate at 1000 GeV. The extra proton energy will be used either to increase photon intensity in the 70-140 GeV range or to double the photon energy. The latter would involve no modification of the electron beam which is capable of 300 GeV."[169] When this suggestion in the proposal is compared with the actual history of E-691, it is clear that the outline for what was later submitted as the proposal for E-691 was already anticipated by the collaboration.[170]

But the first P-516 proposal also laid out a program of study that extended even beyond the boundaries of E-691 stating, "The electron beam can also be used to transport pions into the Tagged Photon Lab. R. Rubinstein notes that although spot sizes will be somewhat larger the intensities are potentially only a factor of about 3 below the P West pion beam. Thus, one can imagine a future proposal to use the spectrometer at the TPL for direct comparison

---

166 For E-516, the PDP 11/55 was used for both data acquisition and data monitoring.

167 The collaboration also added one additional bank of drift chambers to the forward region of the tracking system which enabled them to have finer resolution in resolving the trajectories of particles that penetrated the tracking system.

168 E-691 was formally proposed in February 1981 and not approved by the laboratory until November 1983.

169 The name Energy Doubler was a DOE label attached to what was originally called the Tevatron. See the Proposal to Study Photoproduction of Final States of Mass Above 2.5 GeV with a Magnetic Spectrometer in the Tagged Photon Lab, submitted October 1, 1976, by Jeff Appel, Paul Mantsch, and Tom Nash (Fermilab), Rolly Morrison (University of California, Santa Barbara), and George Luste (University of Toronto), 2-3.

170 I have also established this fact in private communications with Tom Nash, Rolly Morrison, and other members of the E-516/691 collaboration.

of photoproduction and hadron production with systematic errors caused by using different detectors eliminated. This emphasizes the flexibility and long-range benefits to the Proton East program that the spectrometer we propose would bring."[171]

The early assertion that the TPL beamline could be converted to a pion beam, enabling the TPMS to be used for hadronic studies of charm, shows that the original P-516 collaboration outlined the full program for exploiting what was then only a proposed facility. An analysis of the diagrams of the TPMS configurations for experiments E-516, 691, 769, and 791 indicates that the majority of the detector subsystems remained intact and that all substantial changes were simply additions to (or modifications of) the basic structure of the original spectrometer.[172] The core group of physicists from E-516 continued on with E-691, and an analysis of the names on subsequent experimental proposals shows this same core group of physicists constituting the collaborations for experiments E-769 and E-791, mostly with the same institutional affiliations.[173] What did change was converting the primary beamline to transport pions, which enabled the collaboration to study hadroproduction of charmed particles (submitted as the proposal for E-769), and the hadronic decays of charmed particles (submitted as the proposal for E-791).[174]

---

171 See the *Proposal to Study Photoproduction of Final States of Mass Above 2.5 GeV with a Magnetic Spectrometer in the Tagged Photon Lab*, submitted October 1, 1976, 3. The proposal refers to a study performed by R. Rubinstein that supports the feasibility of converting the TPL to a pion beam.

172 The proposal for E-769 (Pion and Kaon Production of Charm and Charmed-Strange States) was submitted to the laboratory in November 1985. The spectrometer configuration was so closely based upon the design of the original E-516 configuration, that the collaboration included a diagram of the TPMS in the E-516 configuration (with the recoil detector) and simply explained the differences in the text. The E-791 proposal did not even contain a copy of the spectrometer design; it had become so well known to the community of physicists at Fermilab.

173 A sample of the core group listed on the proposals is Tom Nash (E-516, 691, 769, 791, Fermilab), Jeff Appel (E-516, 691, 769, 791, Fermilab), Jeff Spaulding (E-516, 691, 769, 791, University of Toronto and later Fermilab), Paul Mantsch (E-516, 691, 769, 791, Fermilab), Steve Bracker (E-516, 691, 769, 791, University of Toronto), and Don Summers (E-516, 769, 791, University of California at Santa Barbara and later the University of Mississippi). Another issue that is beyond the scope of this present study is the relationship between the names listed on the formal correspondence of the experiments to the laboratory (proposals etc.), and the actual level of activity of those people over the course of the experiment.

174 Experiment E-769 was approved in December 1985 and completed in February 1988. Experiment E-791 was formally submitted to the laboratory as a proposal in November of 1987,

This example reveals social and experimental structures and continuities that transcend a single experiment and a sociological framework for analyzing an experimental history that spans fifteen years. Each configuration of the TPMS displays an increasingly complex iteration of the original TPMS yet leaves the fundamental design of the spectrometer intact. Most importantly, the physics goals for the four experiments were already outlined in the original P-516 proposal and reflected a continuing program of measuring the properties of charmed particles using both photoproduction and hadroproduction techniques.

Additional analysis of other experiments performed at Fermilab reveals evidence that experiment "strings" typified Fermilab's entire history, spanning the three laboratory directors, i.e., Robert Wilson, Leon Lederman, and John Peoples. For example, there are similar continuities in the E-82, 226, 383, 425, 486, 584, 617, 731, 773 experiment string that focused on CP violation with a core group led by Valentine Telegdi and Bruce Winstein. Also, the E-8, 440, 495, 555, 620, 619, 756, and 800 experiment string studied neutral (and later charged) hyperon physics, with a core group led by Lee Pondrum, Gerry Bunce, Ken Heller, Tom Devlin, Kam-Biu Luk, Regina Rameika, and Ken Johns. A final example is the E-21A, 356, 616, and 770 experiment string that studied neutrino structure functions and the total neutrino cross section with a core group led by Barry Barish, Frank Sciulli, and Wesley Smith.[175] My preliminary studies indicate that experiment strings were also formed at other laboratories like CERN.[176]

To this point, the physics and detector driven model of social structure has provided a basis for understanding the major continuities that link individual experiments into a string of experiments, but it has not provided a framework that fully describes the deeper motivations for such strings. In other words, I have described what the continuities are, but not *why* these continuities persist through successive iterations of experiments. What are the driving forces that

approved in June 1988, and is currently scheduled for a second running period beginning in January of 1991.

175 Other examples include the E-98, 665 string; the E-1A, 310 string; the E-95, 537, 705, 771 string; the E-87, 358, 400, 401, 402, 687, 831 string; and the E-497, 715, 761 string.

176 In regard to photoproduction experiments at CERN during this same time period there were experiments NA1 (1980 run), NA1 (1983 run), and NA14 as well as experiments WA4, WA57, WA58, and WA69. It is beyond the scope of the present work to describe these examples in detail, but this is part of my ongoing research in this area.

link individual experiments into strings of experiments? Why did this phenomenon emerge at this point in the history of HEP? How can experimental strings be explained within the larger sociological context of laboratory life? More importantly, can they be characterized within the historical context of the development of HEP described in the first section of this paper?

As already indicated, the history of the development of HEP has been characterized by explicit attempts to remove various types of uncertainties that present themselves in the course of scientific practice. Many of the natural uncertainties involved in intermittent cosmic ray particles and small data samples were removed by the design of accelerators. But this thrust experimenters into new sociological uncertainties that for some have proven even more difficult to navigate, namely the sociological problems involved in proton economics and obtaining experimental real estate. The phenomenon of experiment strings is yet another instance of this tendency to remove the physics and sociological uncertainties that emerged within the context laboratories like Fermilab, and the case study of experiments E-516, 691,769, and 791 is a prime example.

In regard to the removal of physics uncertainties, using the tagged photon beamline for all four experiments meant that the properties of the beamline transport system were well known, and this removed many of the uncertainties of starting up a new beamline. This was true independent of the fact that the primary beam was converted from an electron to pion beam.[177] In addition, using the TPMS as a basis for the four experiments made the detector a well-known entity that could be built upon by the collaboration, thus removing the experimental uncertainties of understanding the systematic errors inherent in shaking down a new spectrometer.[178] Because the vast majority of the detector subsystems remained intact throughout all four configurations of the TPMS, much of the reconstruction software for E-516 could be reused with later experiments with simple modifications to the code. This removed many of the uncertainties involved in writing new code for constantly changing detector geometries. In the case of E-691, the experiment's ability to produce

177 As described earlier, the feasibility of converting the secondary beam from electrons to pions had been establish for almost ten years prior to E-769.

178 The advantage of minimalizing the systematic errors of the spectrometer was noted as a justification for viewing the TPMS as a "facility" rather than simply as an "experiment" in the original E-516 proposal. See the *Proposal to Study Photoproduction of Final States of Mass Above 2.5 GeV with a Magnetic Spectrometer in the Tagged Photon Lab*, October 1, 1976, 3.

outstanding physics results in a relatively short amount of time following the end of data taking was a direct result of a conscious decision not to modify the majority of the components of the TPMS from the E-516 configuration.[179] Finally, the program of studying photoproduction then hadroproduction of charmed particles, removed some of the uncertainty of what physics goals to propose to the laboratory. Each series of measurements done by a subsequent experiment was an extrapolation that built upon the previous data and forged ahead into new areas of charmed measurements, accumulating larger and larger samples of fully reconstructed charmed events that yielded increasingly precise measurements of the properties of these particles.[180]

In regard to removing the sociological uncertainties, the ability to maintain possession of the TPL for all four experiments removed many of the uncertainties inherent in experimental real estate and gaining a position at the end of one of the particle beam spigots. In addition, the fact that the TPMS was utilized with only modifications to the original E-516 design removed many of the uncertainties of obtaining the resources needed to build a new spectrometer prior to running an experiment. While the size and complexity of the TPMS drove the original cost of building the E-516 spectrometer, the TPMS proved to be a good investment of laboratory resources, given the physics results it has produced. In addition, the financial investment that the core group institutions made for various TPMS detector subsystems removed many of the uncertainties of obtaining additional financial resources and having the baseline of expertise needed to carry out the subsequent experiments.

The continuity of institutional affiliations also removed many of the uncertainties for the professors and physics departments that needed access to high-energy particle beams to attract and train top-notch graduate students

179 In addition to the evidence of the continuities evidenced in the diagrams of the two experiments, and the statement of this intent in the written text of the E-691 proposal, I confirmed this in a private communication with Rolly Morrison, a member of both experiments.

180 Experiment E-691 accumulated 10,000 fully reconstructed photoproduced charmed events and made the most precise measurements of charmed particle lifetimes in the laboratory. E-769 has accumulated 6,000 fully reconstructed hadronically produced charmed events which was also the largest hadronically produced sample of charmed particles produced in the laboratory. Experiment E-791 plans to accumulate 100,000 fully reconstructed hadronically produced charmed particle events during the current running period at Fermilab, which will surpass even the samples accumulated by photoproduction mechanisms.

in HEP and to publish results regularly.[181] But this experiment string also had a very subtle but real effect upon the Fermilab community's perceptions of the collaboration. With each successive iteration of the TPMS, the collaboration was viewed by their colleagues as a competent group. Their track record at the TPL removed many of the uncertainties about whether they could do what they promised.[182]

By systematically removing these uncertainties, the collaboration mitigated other institutional uncertainties of granting additional resources for each of the experimental progenies. By mitigating these "risk factors," the laboratory would more easily grant the collaboration additional resources and assign them a higher priority than more "risky" experiments who were still in the design, installation, and commissioning phases. Lowering the laboratory management's "risk factor" had a powerful effect on "proton economics." When the Directorate and the Physics Advisory Committee (PAC) decided how to distribute the fixed number of protons produced by the accelerator, successful experiment strings that proposed follow-on studies were more likely to receive the laboratory's support. In fact, at the P-791 proposal presentation to the PAC, the first overhead used by the collaboration had a large picture of a slow-moving turtle on it with a caption that referred to the TPMS as "Old, Slow, and Successful."

Conclusions

The analysis and case study of experiments E-516, 691, 769, and 791 presented above shows how a physics and detector driven model provides insights into the basic social structures in modem HEP experiments at Fermilab and other laboratories. The physics and detector driven model can also be extended to large collider experiments like CDF, where a diagram of the detector and a list of the institutions responsible to design and construct the detector subsystems define the fundamental social structure of the 250 physicists in the

181 Rolly Morrison claims that he was still "putting students through" on the data which he accumulated with E-691 (private communication).

182 The ability to do what they promised was inextricably bound to requests for more and more resources, especially in computing. Thus, while the perception by the HEP community at Fermilab was that they could produce valuable physics results, this heightened the competition with other experiments for laboratory resources. But the ability to "come through" gives an experiment string an advantage over collaborations that cannot appeal to long experimental traditions in support of their requests.

collaboration. The example of collider detectors confirms the fact that the size and complexity of the detector drives the complexity required to "manage" the social structure of the collaboration. With the advent of the next generation of collider detectors like the Solenoidal Detector Collaboration (SDC) and other proposed SSC experiments, physicists must pay even more attention to the problems posed by the social structures of collaborations of over 800 physicists, detectors that cost $700 million (in 1990 dollars) and weigh 10,000 tons, and have 20 million channels of electronics.[183] Even in the early stage of the Expression of Interest (EOI), the SDC collaboration included an explicit statement of the organization and management of the collaboration, complete with a Governance Document which was ratified by a large majority of the collaboration.[184] While it may not be clear how collaborations like the SDC will function throughout the lifetime of the detector, what is clear is that they will probably continue distributing their expertise along the lines drawn in the diagram of the SDC detector, i.e., the detector and physics problems that they know best. This in turn will shape and define the social structure of these mammoth experiments and the processes for designing, building, installing, and operating the experimental apparatus and online and offline computing resources required to reconstruct, analyze and ultimately publish high-energy physics data that continues to push the bleeding-edge of the Standard Model.

The analysis and case study also shows how the emergence of experiment strings was motivated by experimentalists' attempts to avoid the physics and sociological uncertainties produced by the monopoly that laboratory institutions have on the resources for doing HEP. Because some of the experiment strings mentioned above transcend the entire history of Fermilab's research program, and because they seem to emerge so copiously in all areas of HEP research traditions, I believe that the notion of experiment strings, in combination with the recurring attempts of physicists to remove the uncertainties inherent in the institutionalized context of their research, provide a sound sociological and historical foundation for understanding the entire history of the research program at Fermilab throughout all three of the Fermilab Directorates.[185]

183 See the Solenoidal Detector Collaboration Expression of Interest, 24 May 1990.

184 See the Solenoidal Detector Collaboration Expression of Interest, 24 May 1990, 93 ff.

185 This is based upon my ongoing research on the history of the physics program in the Lederman era which is being done under the auspices of the Fermilab History Collaboration.

In the 1980s, the Advanced Computer Program (ACP) at Fermilab was a pioneering, highly cost-effective parallel computing system designed for high-energy physics. It used modular, single-board microprocessors (initially Motorola 68020), to provide massive computing power for data reconstruction, often equaling multiple mainframe computers at a fraction of the cost. For example, a 53-processor ACP farm had demonstrated performance near that of 6 CDC Cyber 175 mainframe computers. Originally developed by Tom Nash from the experiment E-516 and later as Head of the Fermilab Computing Division, the ACP allowed Fermilab to process data from experiments, such as the Tagged Photon Charm Production (E691), significantly faster than traditional, more expensive mainframe systems. In the 1990s, the project led to the development of the ACP Multi-Array Processor System (ACPMAPS) for lattice gauge theory and contributed to the adoption of Linux-based computing farms.

4

Experiment Strings: A Unifying Device that Emerges from the Practice of Science at Fermilab[186]

Introduction

For all the empirical certainty and mathematical rigor of experimental high-energy physics (HEP), its sociology remains somewhat of an enigma to those outside this scientific community. Writing a sociological account of HEP demands that experiments be studied within the institutional contexts where they are performed and not as closed systems divorced from the dynamics of laboratory activities. In addition, social factors cannot be separated from the "experimental" aspects of HEP—detectors, particle beams, and computing architectures. By analogy with the approach taken by Human Factors engineers,

186 This article was submitted to the journal, *Social Studies of Science* in March of 1990, and subsequently published by Fermi National Accelerator Laboratory as Fermilab-Pub-91/63. An earlier version of the article was published by Fermi National Accelerator Laboratory as Fermilab-TM-1706 on December 12, 1990. This work is part of my PhD dissertation research at the University of Chicago, under the direction of Dan Garber and William Wimsatt of the Philosophy Department and Bruce Winstein of the Enrico Fermi Institute. This research was made possible because of the help of the members of the E-516, 691, 769, and 791 collaborations (Jeff Appel, Steve Bracker, Bruce Denby, Rolly Morrison, Tom Nash, Ken Stanfield, and Don Summers) who in interviews and/or by reading drafts provided guidance on the technical details of the paper. I would like to thank my Fermilab colleagues, Anthony Malensek, Jorge Morfin, who read multiple drafts of the paper and made many helpful comments. I would also like to thank G. Fraser, Editor, *CERN Courier* for providing valuable insights from the CERN perspective and Frederick Nebeker (Rutgers, The State University of New Jersey) who provided helpful comments from a historical perspective. Special thanks go to Peter Galison (Stanford University) whose comments helped to strengthen a number of my arguments. Finally, I would like to thank my colleagues Lillian Hoddeson, Catherine Westfall, and Adrienne Kolb for helpful comments on multiple drafts under the auspices of the Fermilab History Collaboration. Fermi National Accelerator Laboratory (Fermilab) is operated by Universities Research Association Inc., for the United States Department of Energy.

those who would study the sociology of experimental HEP must approach it as a people-experiment process.[187] Also, accounts that claim to characterize the nature of experimental practice must focus primarily on experiments not the conceptual development of the theoretical models of HEP. Also, when experimental case studies are used as the basis of accounts, they should be representative of the majority of HEP experiments within a defined time period. Accounts that focus solely on the limited set of so-called "discovery" experiments help to perpetuate the misguided image of high-energy physics research being dominated by a continuing series of crucial discoveries. The account must also be sufficiently fine-grained to describe what collaborations of experimentalists actually do but anchored to the larger institutional context of laboratory life. Writing a sociological account also demands the identification of "unifying devices" that help to organize the data that are presented. But such devices should emerge from the actual practice of experimentalists and not be taxonomies that are artificially imposed on the data.[188] Finally, the sociological researcher should take the views of the experimental researcher seriously, but not uncritically, subjecting them to the constraints described above.

In this article, I present a case study of four high-energy physics experiments performed at Fermilab between 1976 and 1991. I begin by describing the institutional context of Fermilab in terms of a number of distinct, yet interrelated, resource "economies" within which experimental collaborations

---

187 The major objective of Human Factors Engineering (Ergonomics) is the study, design, and development of systems in terms of the capabilities and limitations of the people who operate, maintain, and manage them. My point here is that much like Human Factors engineers study systems with the assumption that the human organism is a crucial element which must be integrated into an overall people-system design, social factors in scientific activities must be viewed as part of the experimental process itself. For details on Human Factors Engineering see B.H. Kantowitz and R.D. Sorkin, *Human Factors: Understanding People-System Relationships* (New York: John Wiley & Sons, 1983), E. J. McCormick and M.S. Sanders, *Human Factors in Engineering and Design*, 5th ed. (New York: McGraw-Hill, 1982), and E. Grandjean, *Fitting the Task to the Man*, 3rd ed. (London: Taylor & Francis Ltd., 1980).

188 The importance of identifying "unifying devices" was recently pointed out by Lillian Hoddeson in a review of J.L. Heilbron's and R. W. Seidel's book, *Lawrence and His Laboratory: A History of the Lawrence Berkeley Laboratory*, vol. 1 (Berkeley, CA: University of California Press, 1989). Hoddeson explains how "Writers of institutional history are typically plagued by too many achievements, too many structures, too many events, and too many people to analyze or even identify carefully. This overabundance of content calls for the invention of unifying devices to prevent disintegration of the histories into kaleidoscopic chronicles." See Lillian Hoddeson, "Roots of Big Science", in *London Times Higher Education Supplement* (London: The London Times, Summer, 1990).

must trade for the resources needed to perform experiments. I then show how the case study suggests a physics and detector driven model of the social structure of collaborations. I also define the experimental process as a people-spectrometer system in which social factors are a crucial part of all phases of the experiment from detector design through the publication of physics results. I try to demonstrate how the physics and detector driven model can be extended to describe "strings" of experiments which have a complex and definable social structure that spans a 15-year period. In addition to describing the copiousness of experiment strings at Fermilab, I suggest that they are motivated by experimentalists' attempts to avoid the physics uncertainties involved in building new experimental detectors, the sociological uncertainties of securing resources from the laboratory, and problems of forming new collaborations to perform follow-up experiments.[189] Finally, I suggest that experiment strings are unifying devices that emerge from the actual practice of HEP and are not ad hoc taxonomies that are imposed upon experimental activities.

Resource Economics at Fermilab

Presently, Fermilab operates the highest energy particle accelerator in the world. The laboratory's five-stage accelerator complex produces 800 billion electron volt (GeV) proton beams which can be used to produce different types of particle beams by colliding them on fixed targets and selecting secondary particles from the collision products.[190] Currently, there is no other accelerator laboratory to which experimentalists can go to obtain particle beams of this high an energy and, consequently, competition for use of these beams is intense.[191] In order to gain access to a particle beam, experimentalists must navigate a

189 In her recent comments for the Panel on Big Science at the 1990 Annual Meeting of the History of Science Society in Seattle, Lillian Hoddeson pointed out that the large investments that collaborations make in their experimental apparatus and the professional relationships they form with members of the group are other important motivations for experiment strings.

190 An electron volt (eV) is the unit of measurement used to describe the acceleration of charged particles like protons and electrons through a one-volt potential. Fermilab's accelerators can also be used to produce beams of protons and antiprotons which are circulated and then collided together in the center of large collider detectors, but I will limit the majority of my discussion to fixed-target counter experiments performed at Fermilab.

191 For example, over the 22-year history of the laboratory, 817 experiments have been proposed by collaborations of experimentalists, but only 345 have been selected, begun, and completed. See Roy Rubinstein (ed), *Fermilab Research Program 1990 Workbook* (Batavia, IL: Fermi National Accelerator Laboratory, 1990), 2. Totals are as of April 1990.

number of interrelated technical and sociological uncertainties within a well-defined institutional structure headed by a single scientist, the Director, who has ultimate authority in all matters scientific and otherwise.[192] Proposed experiments are subject to a complex review process that is embedded in the social structure of the laboratory.[193] While the technological and scientific factors associated with proposing an experiment are more quantitative, the social factors involved in the process of beginning, running, and ending an experiment are much less quantitative and are more difficult for experimentalists to control systematically. While descriptions of these social factors find little or no place in the written accounts of high-energy physicists, for those who are familiar with the actual practice of laboratory life it is clear that they are at least as important as anything "experimental."[194]

The increasing high cost of building larger, higher energy, accelerators and increasingly large and complex experimental detectors are important factors in the overall development of HEP. The financial investment in Fermilab during its 22-year history is over two billion dollars, with a typical fixed target detector costing between three and five million dollars.[195] The magnitude of the financial investment and the fact that these funds are provided primarily by the United States Department of Energy, also introduces a number of political

192 The Director appoints a Physics Advisory Committee (PAC) composed of scientists from universities and other high-energy physics laboratories to advise him on the nature, scope, and priorities of the physics program. This is similar to the organizational structure of CERN. See the *CERN User's Guide* (Geneva, Switzerland: CERN, 1988) for details.

193 For a description of the proposal and review procedure required by the laboratory see the *Fermilab Procedures for Experiments* (Batavia, IL: Fermi National Accelerator Laboratory, 1991). CERN has similar proposal and review processes which it requires of experimental proposals submitted to the laboratory. See the *CERN User's Guide* (Geneva, Switzerland: CERN, 1988) for details.

194 My own familiarity with the actual practice of laboratory life at Fermilab is based upon ten years of observations as a member of the Fermilab staff, most recently in the Fermilab Directorate. Also, I will limit the scope of this article by not discussing the philosophical issues associated with anti-realism or the reduction of the theoretical and experimental aspects of scientific practice in high-energy physics to purely "social interests." For an example of this type of "social constructionist" account see Andrew Pickering, *Constructing Quarks: A Sociological History of Particle Physics* (Chicago: The University of Chicago Press, 1984). For an account which offers a number of cogent arguments against the social constructionist view, see Peter Galison, *How Experiments End* (Chicago: The University of Chicago Press, 1987), 10-13.

195 Fermilab typically operated 15 or more fixed-target experiments simultaneously, and alternates fixed-target operation with two large collider detectors (The Collider Detector at Fermilab, CDF and the D0 detector) which cost over $60 million each.

problems when securing financial resources. While these financial factors are crucial to the study of the development of "big science," I focus on three less understood, and interrelated "economies," (proton economics, experimental real estate, and physicist economics), which are based upon a number of definable "commodities" (protons, experimental halls, and physics expertise).[196] These economies are of course predicated on the assumption that financial resources can be secured by Fermilab, but the analysis reveals a complex social system within which collaborations of high-energy physicists must trade in order to participate in the process of generating scientific knowledge about the subnuclear structure of the physical world.[197]

Within the economy of proton economics, protons are the commodity needed to perform HEP experiments and for which collaborations of

---

196 The phrase "Big Science" was originally coined by Alvin Weinberg, see Alvin M. Weinberg, *Reflections on Big Science* (Cambridge, MA: MIT Press), 1967. Since that time, a number of historians and sociologists of science have made extensive studies on Big Science and the development of particle physics a major focus of their research activities. For example see L. M. Brown and L. Hoddeson (eds.), *The Birth of Particle Physics* (New York: Cambridge University Press, 1986); L. M. Brown, M. Dresden, and L. Hoddeson, "Pions to Quarks: Particle Physics in the 1950s" in L. M. Brown, M. Dresden, and L. Hoddeson (eds.) *Pions to Quarks; Particle Physics in the 1950s* (New York: Cambridge University Press, 1989), p 13 ff; J. L. Heilbron and R. W. Seidel, *Lawrence and His Laboratory; A History of the Lawrence Berkeley Laboratory*, vol. 1 (Berkeley, CA: University of California Press, 1989); A. Pickering, *Constructing Quarks: A Sociological History of Particle Physics* (Chicago: The University of Chicago Press, 1984); P. Galison, *How Experiments End* (Chicago: The University of Chicago Press, 1987); Peter Galison, "Bubbles, Sparks, and the Postwar Laboratory," in L. M. Brown, M. Dresden, and L. Hoddeson (eds.) *Pions to Quarks; Particle Physics in the 1950s* (New York: Cambridge University Press, 1989); D. Kevles, *The Physicists: The History of a Scientific Community in Modern America* (New York: Knopf, 1978); A. Hermann, J. Krige, U. Mersits, and D. Pestre, *History of CERN*, vol. 1 (New York: Elsevier, 1987); A. Hermann, J. Krige, U. Mersits, and D. Pestre, with L. Weiss, *History of CERN*, vol. 2 (Amsterdam: North Holland, 1990); A. Needel, "Nuclear Reactors and the Founding of Brookhaven National Laboratory," in *Historical Studies in the Physical Sciences*, 14 (1983), 93-122; S. Traweek, *Beamtimes and Lifetimes; The World of High Energy Physicists* (Cambridge, MA: Harvard University Press, 1988); B. Latour and S. Woolgar, *Laboratory Life; The Construction of Scientific Facts* (Princeton: Princeton University Press, 1986); B. Latour, *Science in Action* (Cambridge, MA: Harvard University Press, 1987); Michael Riordan, *The Hunting of the Quark* (New York: Simon and Schuster, 1987); S. Schweber, "Some Reflections on the History of Particle Physics in the 1950s," in L. M. Brown, M. Dresden, and L. Hoddeson (eds.) *Pions to Quarks; Particle Physics in the 1950s* (New York: Cambridge University Press, 1989); and C. Westfall, "Fermilab: Founding the First U.S. 'Truly National Laboratory'," in F.A.J.L. James (ed), *The Development of the Laboratory: Essays on the Place of Experiment in Industrial Civilization* (London: Macmillan Press, 1989), 184-217; C. Westfall, "The Site for Fermilab," in *Physics Today*, 42, January 1989, 44-52.

197 The annual Fermilab budget for operating and equipment expenses is currently about one third of the entire high-energy physics budget in the United States, or about $200 million.

experimentalists must compete when submitting experimental proposals.[198] Competition for protons is intense because the magnitude of the overall proton economy is limited by the accelerator's ability to produce a given number of protons (beam intensity). Sociological and beam-management parameters arise when the laboratory Directorate must decide how to utilize the protons, choosing which types of particle beams to produce in view of the submitted proposals.[199] But beam management decisions are further constrained by the cross section for secondary and tertiary beam production.[200] For example, because the cross section for neutrino production is 10^{-36} cm^2 and the pion cross section is 10^{-27} cm^2 the decision to approve experiments that use incident beams of neutrinos is already a major decision that affects proton economics.[201] A neutrino beam is much more costly than a pion beam in terms of the number of protons needed to produce it, especially if a large sample of particle events is necessary for an experiment to achieve its goals.

Proton economics and beam-management decisions are also constrained by the cross sections for particle production in the experimental target. Prior to proposing an experiment, the collaboration performs a Monte Carlo study of the number of potential events to be detected in the experimental spectrometer with a given secondary beam intensity. Knowing the secondary beam intensity, the repetition rate for beam extraction from the accelerator, the nuclear properties of the target, the geometric acceptance of the spectrometer, and the cross section for the interaction to be studied, the collaboration can calculate a

198 I refer to "proton" economics only because I have limited the scope of this article to Fermilab. For example, at the Stanford Linear Accelerator Center (SLAC) where electrons and positrons are accelerated for experiments, one would refer to "electron/positron" economics with the commodities being "electrons and positrons." One could extend this notion to all accelerator facilities, independent of particle type, by referring to "beam" economics, with the generic commodity of particle beams.

199 At Fermilab, the Director appoints a Program Planning Office in the Directorate which oversees beam-management and other issues associated with performing and scheduling experiments. Similar organizational structures exist at CERN. See the *CERN User's Guide* (Geneva, Switzerland: CERN, 1988) for details.

200 When the proton beam is directed toward a fixed target, the protons interact with the nuclei in the target in a variety of ways producing many different types of secondary particle products. The probability that a particular type of secondary particle will be produced is normally expressed in terms of the cross section (σ) per nucleus for each type of interaction. The cross section is traditionally expressed in units of area measured in units of the "barn" 10^{-28} m^2).

201 The pion cross section is roughly constant for energies above two GeV at about 40 millibarns. The neutrino cross section is not constant but is linearly proportional to the energy. For Fermilab, a reasonable neutrino energy to use is 100 GeV, which would give a cross section of about 0.7 picobarns.

reaction rate for that particle event type. The experiment must also determine the number of events that will provide adequate statistics in their data sample, given the size of data samples accumulated by previous experiments. When writing the proposal, the collaboration begins by calculating the size of the event sample and works back through the process to determine the number of incident protons needed on the primary production target. Upon completing these calculations, they include this number as a "beam request" in the experimental proposal they submit to the Fermilab Directorate.[202]

The second type of economy is experimental real estate, the commodity of which is obtaining possession of an experimental hall at the end of one of the particle beamlines to house the collaboration's apparatus. Given the size and complexity of HEP detectors and the long lead times needed to assemble and operate them, then analyze the data they accumulate, collaborations of physicists who are given beam time move into the hall with the explicit goal of performing that experiment, and the implicit goal of not moving out. The detector configurations are often based on the use of general-purpose equipment provided by the laboratory, with only portions of the spectrometer being built by the collaborating universities. Often, an apparatus can be modified to do a series of different experiments.[203]

Collaborations obtain experimental real estate in a number of ways, all of which aim at "holding their place in line" and preventing competing collaborations from gaining control of a specific piece of real estate. For example, incumbent experiments try to convince the laboratory of the importance of their proposed physics measurements with the goal of securing more laboratory resources for a follow-up experiment. The resources are more than just beam time, but usually involve the resources needed to upgrade the experimental detectors themselves. If they succeed, they can leave the major components of their detector in place, upgrade portions of the apparatus, and gain additional

---

202 For example, at the time of this writing, a rough summation of all the maximum beam requests for experiments proposing to run during the 1993 fixed-target run at Fermilab is about $5X10^{13}$ protons. The accelerator is currently capable of producing about $1.5X10^{13}$ protons, a factor of three less than current requests. The constraints of proton economics will limit the total number of experiments which will actually be approved to run by the Director or the scheduling of experimental running periods.

203 The practice of using Fermilab provided, general purpose equipment, became more common as experimental configuration increased in cost, size, and complexity. An early example that I will focus on in the case study is the Tagged Photon Magnetic Spectrometer, which is located at the Tagged Photon Lab.

access to the beam. If the above strategy fails and other experimental groups threaten to displace them by submitting a proposal that uses that beamline and experimental hall, the next best trading strategy is for the incumbent to try to convince the laboratory (and the competition) to use portions of their detectors (something which the competing group often argues against). Doing this enables the incumbent to partially hold his place in line and increases the probability that the laboratory will allow them to do subsequent experiments. In generating their arguments against giving up the experimental real estate, the incumbent collaboration uses the data from the previous run and the promise of new results to show that the experiment is "more important" to the study of HEP than the proposed results of the experiment seeking to displace them. Sometimes this strategy works, and other times it fails because, within the economy of experimental real estate, the laboratory Director must ultimately decide which experiments gain access to a particular piece of experimental real estate.

The third type of economy is physicist economics, with the commodity being the physicists themselves and their specific areas of expertise. Larger and more complex detector configurations demand that collaborations contain larger numbers of physicists having a distribution of experimental expertise. The number of high-energy physicists that can commit themselves to perform experiments at Fermilab is constrained by the total number of physicists available at a given point in time and the rate at which PhD graduates are produced. Consequently, physics expertise had become a valuable commodity within the resource economies of Fermilab.[204] In addition to defining the physics measurements to be made and the detector design, one of the most important factors of whether an experimental proposal will be approved by the Fermilab Directorate, is the "people design" of the collaboration. Within the scientific proposal, the collaboration must demonstrate to the laboratory that it has a sufficient number of physicists committed to carrying the experiment

---

204 The issue of physicist economics is a crucial part of evaluating the direction of presently operating and future accelerators (like the Superconducting Super Collider) with their associated experiments. For example, A recent study of the HEP research program for the 1990s performed by the High-Energy Physics Advisory Panel, under the auspices of the United States Department of Energy, included a detailed demographic study of "manpower considerations" during the time period under study. See the *HEPAP Subpanel on the U.S. High Energy Physics Research Program for the 1990s*, U.S. Department of Energy Office of Energy Research Division of High Energy Physics, DOE/ER-0453P, April 1990, pp 68 ff.

through to the final stage of publication.[205] This also means that an adequate number of graduate students will be sponsored by the university-based contingencies of the collaboration. The people design described in the proposal must also show that the collaboration has the appropriate distribution of expertise needed to design, fabricate, install, commission, and operate the apparatus. In addition, it must account for the expertise needed to develop the computing systems and the software programs used to reconstruct and analyze the particle events recorded with the detector.

Collaborations that propose experiments are often formed around a core group of physicists who have specific physics interests. One of the most common ways members of the core group recruit collaboration members is by making presentations to their colleagues in the hope of attracting them to join the experiment. Most times the presentation consists of a description of the physics measurements to be made, a technical description of the detector subsystems, and a "shopping list" of "job openings," which normally involves assuming responsibility for one or more of the detector subsystems of the overall experimental apparatus.[206] The notion of the "people design" is important because it is a heuristic, which points to a method for characterizing the basic social structure of HEP experiments and the higher-level social structures that are required by the laboratory Directorate.

A Physics and Detector Driven Sociology: A Case Study

The history of the development of the detectors used for HEP moved along two definable but orthogonal axes (size and complexity), both directly resulting from the physics measurements being made. Along the first axis, detectors increased in size, but not necessarily in complexity. An example highlighted by Peter Galison was the development of the table-top sized spark chamber by J. Cronin and G. Renninger in 1960, and its expansive extrapolation by Melvin

205 In addition to teaching responsibilities, university-based physicists often divide their research time between more than one experimental activity, while Fermilab-based physicists can devote 25-50 percent of their time to performing experimental research, which is considered a part of their laboratory responsibilities. Consequently, the number of physicists committed to an experiment is often expressed in terms of the number of full-time equivalents (FTEs), with experimentalists devoting only a portion of their research time to an experiment. In other words, five physicists who devote 20 percent of their time to a single experiment are counted as a single FTE for the collaboration.

206 Another common way that experimentalists recruit collaborators is by individually contacting colleagues who they know have technical expertise in a particular area.

Schwartz and Leon Lederman into a detector that was large enough to use surplus naval cruiser deck plates weighing between two and three thousand tons.[207] The scale of this type of counter detector increased rapidly. Within ten years, experimentalists had moved from the table-top to the mammoth spark chambers used by the Fermilab based E-1A experiment. What is important to note about the scale of the E-1A detector is that its size was driven by the fact that the E-1A collaboration was searching for experimental evidence of neutral currents using neutrino interactions.[208] It was the cross section for neutrino interaction that necessitated a target mass measured in tons and the increased requirements for shielding the fiducial volume of the detector from background events. Consequently, the design of such experiments was inextricably tied to the physical properties of the neutrino cross section and the cross sections of the hadrons and leptons that composed the background for the events E-1A chose to study. But because the signature for detecting neutral currents only involved detecting the absence or presence of a muon track protruding from the cluster of hadrons produced in the incident neutrino collision, the E-1A detector (for all its size) was a relatively uncomplex detector, having only 46 spark chambers. Growth in the size of detectors also meant an increase in the cost of experiments like E-1A, which made collaborations more dependent on Fermilab for the financial resources needed to construct these mammoth experimental configurations.

Along the orthogonal axis of complexity, many varieties of detector types were combined to form increasingly complex spectrometers composed of a variety of detector subsystems. These spectrometers enabled experimentalists to record more complex physical interactions and measure an increased number of physical parameters simultaneously. The Tagged Photon Magnetic

---

207 This detector was used in 1961 to demonstrate that there were two distinct types of neutrinos, the muon and electron neutrinos. L. Lederman, J. Steinberger, and M. Schwartz later shared the 1988 Nobel Prize in Physics for this discovery. See Peter Galison, "Bubbles, Sparks, and the Postwar Laboratory," in Laurie M. Brown, Max Dresden, and Lillian Hoddeson (eds.) *Pions to Quarks: Particle Physics in the 1950s* (New York: Cambridge University Press, 1989), 235-237.

208 The issues associated with the first axis of detector development, "The Scale of High Energy Physics," are discussed by Peter Galison in reference to experiment E-1A and the Gargamelle bubble chamber at CERN. Galison also provides a detailed account of the E-1A experiment and carefully describes the collaboration's search for evidence of neutral currents. I will not reiterate these details here. See Peter Galison, *How Experiments End* (Chicago: The University of Chicago Press, 1987), 197 ff and Peter Galison, "How the First Neutral Current Experiments Ended" in *Rev. of Mod. Phys.*, 55 (1983): 477-509.

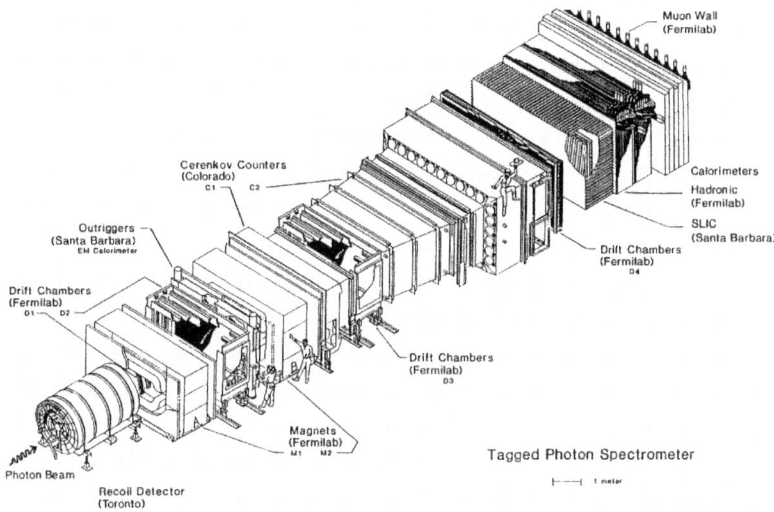

Figure 1: The Tagged Photon Magnetic Spectrometer.

Spectrometer (TPMS) used for experiment E-516 at Fermilab was an early example of a spectrometer that grew substantially along both axes simultaneously, being large in size and very complex.[209] The goal of experiment E-516 was to do precise studies on charmed particles using a photon beam as incident on an experimental target of liquid hydrogen.[210] When the incident photon beam interacted in the experimental target, it produced numerous secondary particles, some of which were charmed particles that would subsequently decay into other types of particles. The Tagged Photon Magnetic Spectrometer was

209 See the *Tagged Photon Magnetic Spectrometer; Facility Design Report*, May 9, 1977, for the technical details of the spectrometer's design. Other less sophisticated examples at Fermilab were the Multi Particle Spectrometer used for experiment E-557 and the spectrometer used for experiment E-400 in the Broad Band Photon Beam Line.

210 Charmed particles contain at least one charmed quark. They were originally discovered independently in 1974 by S. Ting et al at Brookhaven National Laboratory and B. Richter et al at the Stanford Linear Accelerator Center using different experimental techniques. Ting and Richter shared the 1975 Nobel Prize in Physics for the discovery. For an account of this discovery, see Michael Riordan, *The Hunting of the Quark* (New York: Simon and Schuster, 1987). The formal name of E-516's original proposal was "Proposal to Study Photoproduction of Final States of Mass Above 2.5 GeV with a Magnetic Spectrometer in the Tagged Photon Lab" which was submitted to the Fermilab Directorate on October 1, 1976 by J. Appel, P. Mantsch, and T. Nash (Fermilab), R. J. Morrison (University of California, Santa Barbara), and G. Luste (University of Toronto).

designed to detect the 8–12 decay products and allow the E-516 collaboration to record, reconstruct, and analyze properties like the charmed particle production cross sections and particle mass.

As shown in Figure 1, the E-516 spectrometer was an array of individual detector subsystems designed to function as a unit, with events selected at a higher level by a specially designed computing architecture.[211] Because the decay products of charmed particles were more numerous than simply detecting the absence or presence of a muon track for E-1A, the number of data measurements needed to record and reconstruct the multiple final states demanded that the spectrometer be a significantly more complex apparatus. The E-516 spectrometer was divided into six detector regions: a) the experimental target/recoil detector, b) the tracking system, c) the Cerenkov counters, d) the electromagnetic calorimetry, which consisted of the segmented liquid scintillation shower counter (SLIC) and the outriggers, e) the hadron calorimeter, and f) the muon system.[212] The complexity of the particle interactions to be detected was reflected in the complexity of the spectrometer which had 7,000 interaction points distributed over the six detector regions.[213] Each interaction point was connected to a channel of electronics and fed into the trigger processor and finally the online computing system. The design of the E-516 spectrometer demonstrates the relationship between the complexity of the physics interactions to be measured, and the complexity of the detectors needed to detect and measure those interactions. It also shows that the cost of building such spectrometers

211 Experiment E-516 used a computerized trigger processor which allowed the collaboration to trigger the spectrometer on interesting events by programming the trigger requirements into the computer. It was specially designed by T. Nash and S. Bracker and was described in the May 1983, issue of *Physics Today*.

212 The target/recoil detector was used primarily to trigger the spectrometer, the tracking system was used to analyze the trajectories and momenta of charged particles that penetrated the apparatus, the Cerenkov counters were used to identify particle type by measuring particle mass, the electromagnetic calorimeters (SLIC and outriggers) measured the energy of photons and electrons, the hadron calorimeter was used to measure the energy of uncharged hadrons, and the muon system was used to measure the presence and trajectories of muons that penetrated the most forward region of the spectrometer. For the technical details of the detector subsystems see the *Tagged Photon Magnetic Spectrometer; Facility Design Report*, May 9, 1977.

213 Physicists normally refer to an "interaction" point as a point at which particles interact in the experimental target. What I am referring to are points within the detector subsystems where the decay products from particles produced in the experimental target interact, for example, the cell of a drift chamber in the tracking system, or a wire in a proportional wire chamber in the target/recoil detector.

substantially increased by virtue of both size and complexity, because obtaining more interaction points and channels of electronics meant obtaining more money.[214] When spectrometers became as large and complex as the E-516 spectrometer, the institutions that composed the collaboration not only became more dependent on the laboratory for financial resources, they also became more interdependent upon one another for bearing their share of the cost of the experiment. The comparison between E-1A and experiment E-516 suggests that the nature of the physics measurements made by these experiments powerfully drove the physical architecture of the detector design. For those familiar with the experimental practice of HEP, this claim ought not be surprising.

But what may be surprising is the claim that the physics driven explanation of detector designs does not stop at the detectors but extends to the social structure of the collaboration itself.[215] In other words, the "people design" of

214 A rough rule of thumb for estimating the expense of complex spectrometers is to quantify cost in terms of dollars per channel of electronics. K. Stanfield, who designed the drift chambers for the E-516 tracking system, claims that one of the major factors in determining the number of wires/channels for the tracking system was taking the dollar amount of the E-516 budget for tracking and dividing by the cost per wire and channel of electronics (private communication).

215 Discussion about whether high-energy physics is "driven" by theoretical or experimental developments, while important to the conceptual development of HEP, would lead far beyond the scope of offering a fine-grained account of the experimental activities in my case study. What is at stake here philosophically is understanding the role that experiment plays in the overall development of HEP. These issues are typically discussed in terms of the Positivist and post-Positivist tendency to make a clear distinction between observation and theory and the Kuhnian and post-Kuhnian tendency to make little or no distinction between observations and theoretical or sociological background assumptions. But whether the experimentalists of E-516 believed that "physics" was driven by theoretical developments, social interests, or experimental advances, I claim that the account presented here still obtains. Consequently, I will continue to abstain from discussing these philosophical issues in what follows. For discussions on the Positivist view see A.J. Ayer (ed), *Logical Positivism* (New York: The Free Press, 1959). For discussions on a recent reformulation of Positivist notions see Bas C. van Fraassen, *The Scientific Image* (Oxford: Clarendon Press, 1980) and Paul M. Churchland and Clifford A. Hooker (eds), *Images of Science: Essays on Realism and Empiricism*, with a Reply from Bas C. van Fraassen (Chicago: The University of Chicago Press, 1985). For a discussion of the Kuhnian view see Thomas S. Kuhn, *The Structure of Scientific Revolutions*, 2nd ed. enlarged (Chicago: The University of Chicago Press, 1970). For a discussion of the post-Kuhnian "social constructionist" view see Andrew Pickering, *Constructing Quarks: A Sociological History of Particle Physics* (Chicago: The University of Chicago Press, 1984) and B. Latour, *Science in Action* (Cambridge, MA: Harvard University Press, 1987). For discussions on the importance of the role of experiment in the development of science in a non-Positivist/post-Positivist and non-Kuhnian/post-Kuhnian framework see Ian Hacking, *Representing and Intervening* (New York: Cambridge University Press, 1987) and Peter Galison, *How Experiments End* (Chicago: The University of Chicago Press, 1987), 6-13.

the E-516 collaboration was built around the physical structure of the E-516 spectrometer. In the initial organization of the collaboration, each of the six regions of the spectrometer became the domain of a particular institutional group in the collaboration and represented that group's hardware contribution to the overall experiment. The recoil detector was built by the University of Toronto, the SLIC and outriggers were built by the University of California at Santa Barbara, the Cerenkov counters were built by the University of Colorado at Boulder, and the tracking system, hadron calorimeter, and muon system were built by Fermilab scientists (see Figure 1).[216]

The basic social structure of experiment E-516 emerges from the drawing of the E-516 spectrometer when the universities and institutions responsible for the design and construction of the detector subsystems are identified. Add to this the fact that each of these institutions had an institutional representative, and the basic social structure of E-516 begins to take shape. The complexity of the physics interactions determined the physical structure of the spectrometer, and the people who planned to carry out the experiment organized themselves around this structure by institution and area of expertise.[217] Because the E-516 spectrometer was a complex array of detector subsystems, it was carved up into smaller, more manageable pieces, which suggests how the phenomenon of technical specialization for experimenters emerged with the advent of large and complex experimental configurations. Those who are interested in characterizing the basic social structure of experiments like E-516 must begin their analysis with a diagram of the apparatus itself. In a well-defined and predictable way, the increased complexity of the physics measurements led to an increased complexity in spectrometer designs, which in turn meant more complex social structures for collaborations. The descriptive ability of the physics and detector driven model can be extended even further to describe most aspects of the E-516 experiment up to and including the data analysis and publication phases.

216 The trigger processor and the online computing systems were collaborative efforts between Fermilab and the University of Toronto contingents of the collaboration.

217 It is interesting to note that the social structure of Fermilab itself is organized in a similar way. The Fermilab organizational chart shows that the 2,200 people employed by the laboratory are distributed around the accelerators, devices, and support systems needed to produce the particle beams for experiments. I attribute this largely to the fact that the upper-level management positions at Fermilab are held primarily by high-energy physicists who tend to define the laboratory's social structure much like an experimental collaboration would.

In the design, construction, and installation phases, the technical and financial demands of getting the spectrometer built and operating forced the collaboration members to focus most of their attention on that region of the spectrometer for which they were responsible. Because many of the detector subsystems were constructed at home institutions then shipped to Fermilab for installation, there tended to be less social interaction between collaboration members in the earlier phases of the experiment.[218] The amount of social interaction significantly increased when they came together as a group at the laboratory during the installation phase of the experiment. From this point on, the level of social interaction increased through the actual data taking phases of the experiment. But the E-516 spectrometer had even more profound effects on the nature of the collaboration's social interaction in the later phases of the experiment.

During the actual running of the experiment, the detector-based social taxonomy defined the way in which technical problems and challenges were addressed. Problems in the operation of the segmented liquid scintillation shower counter (SLIC), for example, were within the domain of the University of California, Santa Barbara contingent that designed and constructed it. Problems with the Cerenkov counters were considered the purview of the University of Colorado at Boulder. Various members of the collaboration gained a limited knowledge of the detector subsystems for which they had no direct responsibility. But throughout the operation phase of the experiment, the major repositories of expertise for running the experiment were taxonomized and distributed along the lines shown in the diagram of the E-516 spectrometer.

The phase of the experiment in which the collaboration wrote the computer software programs needed to reconstruct particle events (and the process of reconstructing them) was spread over a three-year period.[219] During the recon-

218 There were of course design meetings and reviews performed both within the collaboration and in the context of the laboratory, but the fact that detector components were constructed at home institutions and subsequently shipped to Fermilab limited the amount of social interaction within the collaboration during the earlier phases of the experiment.

219 The reconstruction and running phases of E-516 overlapped substantially in time. This is not always the case with even large and complex experiments, but depends on the level of software expertise in the collaboration, the "real-world" constraints of getting the hardware portions of the detector up and running properly, the complexity of the physics events and the spectrometer, and the level of competition within the collaboration in regard to which individuals write which software packages. In regard to dates, the first shakedown run of E-516 was in the Summer of 1979, with the first data run beginning in the late fall of 1979. The final data run began in early 1981 and ended at the accelerator

struction phase, there seemed to be a relaxing of the well-defined distribution of hardware expertise, as various members who had responsibility for one specific detector subsystem wrote software reconstruction packages for other regions of the spectrometer. In fact, multiple reconstruction packages were written for some of the detector subsystems.[220] During the reconstruction phase the boundaries that organized the collaboration around the spectrometer's hardware were more easily crossed as one group member attempted to extend his area of expertise to detector subsystems, which were the responsibility of another group in the collaboration. While the collaboration's social structure was anchored in the spectrometer design throughout all phases of an experiment, the example of E-516 shows that these taxonomies became less rigid and less defined in the later phases of the experiment when tasks involved the development and use of software rather than hardware.[221]

In the final phase of the experiment, data analysis and final publication, the collaboration faced a more intimate type of social interaction, which was strongly related to the distribution of expertise around the spectrometer. Problems associated with the final calibration of data recorded with the SLIC were addressed mainly by the University of California, Santa Barbara group that designed and built the detector, while problems with the final calibration of data recorded with the Cerenkov counter were considered within the domain of the University of Colorado, Boulder group.[222] What is interesting sociologically about this experimental phase is that in the same way all of the physical components of the TPMS had to work together as a unit in order to produce the evidence that would support the data in the publications, the human components that were distributed around the detector subsystems were also forced to work together as a unit. In the same way the six spectrometer regions had to work together as a unit in order to analyze the multiple decay products of charmed particles, the collaboration had to work together

---

shutdown in June 1981, with the data reconstruction process continuing into the early part of 1983.

220 For example, there were three separate reconstruction packages generated for the tracking system, and two packages generated for the SLIC.

221 This is partially due to the inability of the collaboration to secure the resources necessary to design and construct multiple hardware detector subsystems. The resources needed to produce software are much easier to obtain because they only involve the time and expertise of the students and physicists on the experiment that are directly under their control and have little or no financial impact.

222 Evidence from the minutes of collaboration analysis meetings during this phase of the experiment supports this claim.

as a people-spectrometer unit to come to agreement on the final form of the publications.[223] While the "human" aspects of individual personalities were manifested throughout all phases of the experiment, they were evidenced more powerfully in the more intimate social negotiation needed to produce physics results.[224]

In addition to the spectrometer-based social structure, there was a higher-level social structure that is important to describe. The Fermilab Directorate requires experimentalists to appoint a spokesperson that acts as the liaison between the laboratory and the collaboration in all aspects of the experiment. The "management" style of the spokesperson is also an important element in the social negotiation that occurs both within the collaboration and between the experimentalists and the Fermilab Directorate. Almost without exception, the spokesperson and the laboratory Directorate are the major players in what might be characterized as the Director-experiment relationship. While the spokesperson's organizational and interpersonal skills are important to the "management" of the collaboration, they are far less important to defining the group's basic social structure. Whether the management style of the spokesperson is autocratic, democratic, bureaucratic, etc., the underlying spectrometer-based social structure is the foundation upon which these higher-level social structures are predicated.[225] For example, shortly after the follow-up experiment to E-516 was approved by the Fermilab Directorate, there was a rotation of spokespersons for the collaboration.[226] But the basic

223 The two major papers from the experiment were submitted for publication in the Fall of 1983. They appeared in the journal *Physical Review Letters* in early 1984. See D.J. Summers et al, "Study of the Decay D0 $\rightarrow K^- \pi^+ \pi^\circ$ in High-Energy Photoproduction," in *Phys. Rev. Lett.* Vol. 52, No. 6, 6 February 1984, pp 410-413, and B. Denby et al, "Inelastic and Elastic Photoproduction of J/psi (3097)", in *Phys. Rev. Lett.* Vol. 52, No. 10, 5 March 1984, 795 ff.

224 The written correspondence between members of the E-516 collaboration shows that the collaboration experienced significant intergroup problems determining when the data results had stabilized and were dependable enough to publish. Galison characterizes this type of negotiation in HEP experiments as occurring along two axes: increasing directness of the measurement and growing stability. See Peter Galison, *How Experiments End* (Chicago: The University of Chicago Press, 1987), 259 ff.

225 The issue of the "management style" of collaboration spokespersons was recently pointed out in reference to the spokespersons for experiments competing for approval to run at the Superconducting Super Collider Laboratory. The management style of two of the spokespersons, S. Ting (L* experiment) and G. Trilling (SDC experiment), are described respectively as "autocratic" and "bureaucratic." See Robert Crease, "Choosing Detectors for the SSC" in *Science*, Vol. 250, 12 December 1990, 1648-1650.

226 T. Nash, spokesperson for E-516, rotated the position of spokesperson with another member of

social structure of the collaboration and the repositories of expertise remained taxonomized around the physical design of the spectrometer even after this change of leadership.[227]

The example of how the detector subsystems working together forced the collaboration to work together as a unit shows how the practice of HEP is an interactive people-spectrometer system from which sociological factors cannot be removed. Descriptions of how scientific knowledge in HEP is produced must be characterized in terms of the experimentalists and their detectors working together to produce the data that constitutes the evidence presented in scientific publications.

This example also reveals another crucial sociological phenomenon that extends the descriptive ability of the physics and detector driven model beyond a single experiment to "strings" of experiments, revealing even more complex, definable, and long-lived social structures.

The Nature of Experiment Strings

Experiment E-516 and its follow-up experiment (E-691) met at a transition-like interface where the first experiment transformed into a second one.[228] Most physicists in HEP understand this transformation process intuitively and can even identify experiment strings that consist of four or five individual experiments. But such intuitions have yet to be documented in a systematic way.[229]

---

the E-516 collaboration, M. Witherell, in order to devote his time to the development of new low-cost high-power parallel processing computers which he had proposed to the laboratory a year earlier. See "A Program for Advanced Electronics Projects at Fermilab" by Tom Nash, May 11, 1982. This proposal eventually developed into the Advanced Computing Project (ACP) under the leadership of Nash.

227 For some Fermilab experiments, like E-665, the role of spokesperson is a fixed time-period (much like the chairman of a university department), after which another member of the collaboration assumes that role.

228 Follow-up experimental proposals which require a change in experimental configuration are once again subjected to the Fermilab review process and, if approved by the laboratory Directorate, are assigned a new experimental number. In this case, the follow-up experiment for E-516 was designated as E-691. Experiment E-691 was formally proposed to the laboratory on February 4, 1981, before the end of the final data run of E-516. Much like E-516, the goal of E-691 was to study the photoproduction of charmed particles using an incident photon beam with the Tagged Photon Magnetic Spectrometer. But experiment E-691 ran with a proton beam of 800 GeV from the newly completed Tevatron accelerator which raised the energy of the incident photon beam to the experiment from about 160 GeV to about 300 GeV.

229 I am aware of similar, but independent, work in process at the Center for the History of Physics of the American Institute of Physics (AIP), headed by J. Warnow-Blewett. While the AIP group's overall

Key to describing this transformation is the ability to characterize the continuities between the individual experiments in the experiment strings. The evidence described below suggests that these continuities are constituted by the physics goals, the spectrometer configuration, the physicists in the core group of E-516, E-691 and their progeny, plus the collaboration's ability to work together as a people-spectrometer unit in a continuing program of physics studies that yields successful, published results.

The configurations of the Tagged Photon Magnetic Spectrometer for E-516 and E-691 were almost identical. The one important difference was the use of a Silicon Microstrip Detector (SMD) instead of the recoil detector and more loosely defined trigger processor assumptions.[230] The SMD allowed E-691 to reconstruct the decay vertices of charmed particles in the upstream portions of the spectrometer and project the tracks of particle decays to the detector subsystems located downstream. In contrast, experiment E-516 triggered the spectrometer with the upstream recoil detector and projected the particle tracks back toward a vertex that could not be as carefully defined by the recoil detector. The use of the Silicon Microstrip Detector was the only major hardware change, revealing a strong spectrometer-based continuity between the two experimental configurations.[231] This modification to the spectrometer was adopted as an additional "job" that the University of California, Santa Barbara element of the collaboration performed, which meant that the spectrometer diagram shown in Figure 1 continued to define the basic social structure of

research goals are to describe multi-institutional collaborations at a number of laboratories, J. Genuth's and F. Nebeker's work discuss the existence of an experiment string at Fermilab. Nebeker's research focuses on experiments E-70, 288, 494, 605, 608, 772, and 789.

230 While both E-516 and E-691 triggered the spectrometer on charmed particles, E-516 used the recoil detector in combination with tightly defined trigger assumptions to look for the charm signature of a recoil proton. E-691 used more loosely defined trigger assumptions with portions of the forward region of the spectrometer to trigger the spectrometer. It is interesting to note that the configuration for E-691 was so similar to the one used for E-516, that the collaboration simply attached a copy of the original *Tagged Photon Magnetic Spectrometer: Facility Design Report* produced in 1977 for E-516 to their proposal and described the differences between the two configurations in the text.

231 The discontinuities between the two configurations included the addition of two banks of drift chambers to the tracking system (which enabled them to have greater redundancy in resolving the trajectories of particles penetrating the spectrometer), and the addition of increased online computing capabilities (using a VAX 11/780 for data monitoring and a PDP11/55 solely for the acquisition of data from the detector subsystems). For E-516, the PDP11/55 was used for both data acquisition and data monitoring.

the collaboration in a predictable way.[232]

The physics goals described in the original proposal of E-516 already included an ambitious physics program for the spectrometer that transcended the studies proposed by the first experiment. The plans for using the spectrometer in the E-691 configuration were already explicitly stated as early as 1976.[233] The proposal states,

> "In fact, although this is by no means a proposal for the Energy Doubler, this spectrometer will be ideal—and unique—for studying photoproduction when the Doubler comes into operation. It is probably the only existing P East facility that will be able to operate at 1000 GeV. The extra proton energy will be used either to increase photon intensity in the 70-140 GeV range or to double the photon energy. The latter would involve no modification of the electron beam which is capable of 300 GeV."[234]

When this statement is compared with the actual history of E-516's progeny, it is clear that the outline for what was later submitted as the proposal for E-691 was already anticipated by the collaboration five years earlier.[235] But the original E-516 proposal also described a program of study that extended beyond the boundaries of a second experiment by stating,

> "The electron beam can also be used to transport pions into the Tagged Photon Lab. Thus, one can imagine a future proposal to use the

232 Responsibility for the design, construction, and installation of the Silicon Microstrip Detector was given to the University of California, Santa Barbara group. The additional online computing was a collaborative effort between the University of Toronto and Fermilab.

233 E-691 was formally proposed in February 1981, but not approved by the laboratory until November 1983.

234 The name "Energy Doubler" described a (then proposed) new ring of superconducting magnets designed to increase the energy of the accelerator proton beam from 400 to 1000 GeV. This accelerator ring eventually became known as the Tevatron, because the design goal was to produce proton beams of 1000 GeV, or one trillion electron volts. See the *Proposal to Study Photoproduction of Final States of Mass Above 2.5 GeV with a Magnetic Spectrometer in the Tagged Photon Lab*, submitted October 1, 1976, by J. Appel, P. Mantsch, and T. Nash (Fermilab), R. J. Morrison (University of California, Santa Barbara), and G. Luste (University of Toronto), 2-3.

235 I have also confirmed this in private communications with T. Nash (E-516 spokesperson), R. J. Morrison, and other members of the E-516/691 collaboration.

> spectrometer at the TPL for direct comparison of photoproduction and hadron production with systematic errors caused by using different detectors eliminated. This emphasizes the flexibility and long-range benefits to the Proton East program that the spectrometer we propose would bring."[236]

The dual statements about converting the beamline to a pion beam and using the spectrometer to do hadronically produced charm studies reveal a plan for exploiting a continuing program of physics measurements with what was at the time only a proposed facility. The primary beam was eventually converted to transport pions enabling the collaboration to study hadroproduction of charmed particles (submitted as the proposal for E-769), and then hadronic decays of charmed particles (submitted as the proposal for E-791).[237] A listing of the names on the proposals for the four experiments reveals that there was a distinct continuity in the core group of physicists who proposed the subsequent experiments.[238] In addition, an analysis of the components of the spectrometer configurations for experiments E-516, 691, 769, and 791

---

236 See the *Proposal to Study Photoproduction of Final States of Mass Above 2.5 GeV with a Magnetic Spectrometer in the Tagged Photon Lab*, submitted October 1, 1976, p 3. The proposal refers to a study performed by R. Rubinstein that supports the feasibility of converting the TPL beamline to a pion beam.

237 The iterations from one experiment to another show how collaborations attempt to "hold their place in line" within the resource economy of experimental real estate. While there were no competing experimental collaborations attempting to displace E-516 during this time period, it is interesting to note that the E-516 collaboration originally obtained the real estate at the Tagged Photon Laboratory by displacing experiment E-152. In attempts to resist this take over, the spokesperson for E-152, C. Heusch, proposed to the Fermilab Directorate that the two experiments share "portions" of the E-152 spectrometer, hoping to partially hold his place in line in the experimental hall. See Heusch to Goldwasser, June 2, 1977. The proposal for experiment E-769 was submitted to the laboratory in November 1985, approved in December 1985 and completed in February 1988, having a substantial time overlap with its predecessor, E-691. The proposal for experiment E-791 was formally submitted to the laboratory in November of 1987, approved in June 1988, and is currently scheduled for a second running period early in 1991.

238 A sample of the core group listed on the proposals is T. Nash (E-516, 691, 769, 791, Fermilab), J. Appel (E-516, 691, 769, 791, Fermilab), W. J. Spalding (E-516, 691, 769, 791, University of Toronto and later Fermilab), P. Mantsch (E-516, 691, 769, 791, Fermilab), S. Bracker (E-516, 691, 769, 791, University of Toronto), and D. Summers (E-516, 769, 791, University of California at Santa Barbara and later the University of Mississippi). Another issue that is beyond the scope of this present study is the relationship between the names listed on the formal correspondence with the laboratory and the actual level of activity of these individuals over the course of the experiment.

shows that the majority of the detector subsystems remained intact and that all substantial changes were simply additions to (or modifications of) the basic structure of the spectrometer.[239]

This case study provides support for a physics and detector driven model, with the first major continuity, the physics goals of the four experiments, already outlined in the original E-516 proposal as a continuing program of measuring the properties of charmed particles using both photoproduction and hadroproduction techniques. Two other major continuities of this experiment string are the spectrometer configuration and the core group of physicists who proposed the experiments. The case study suggests that there are social and experimental continuities that transcend a single experiment and can provide a method for understanding what appears to be more complex social structures and research programs that exist for 15 years.[240] Each experimental configuration in this string displays a more complex iteration of the original spectrometer, which leaves the fundamental design of the detector subsystems largely

239 As with the E-691 proposal, the spectrometer configuration for experiment E-769 was so closely based upon the design of the original E-516 configuration, that the collaboration included a diagram of the E-516 configuration of the spectrometer with the experimental proposal and simply explained the modifications in the text. The E-791 proposal did not even contain a copy of the spectrometer design, it had become so well known to the community of physicists at Fermilab.

240 I have purposely limited my discussion to fixed-target counter experiments and not discussed the experiments performed with large bubble chambers like the Fifteen Foot Bubble Chamber at Fermilab. As a brief comparison with our case study, the major continuity between experiments performed with the Fifteen Foot Bubble Chamber (E-28A, 31A, 45A, 53A, 155, 172, 180, 202, 234, 341, 343, 380, 388, 390, 502, 545, 546, 564, and 632) seems to be the chamber itself. In a less defined way, there were some continuities in the target substances with which the chamber was filled. But the social structures of these collaborations were different from the one described in the E-516 case study. Bubble chamber spokespersons seemed to draw upon the expertise of the international community of bubble chamber physicists each time they formed an experimental group and consequently the collaborations did not exhibit the same type of well-defined core-group structure found in large, complex fixed-target counter experiments. Preliminary studies show that the relatively uncomplex social structure of these collaborations results from the existence of a Fermilab-based Bubble Chamber Department devoted solely to the operation and maintenance of the complex systems of the chamber, independent of the experimental collaborations that used it. When a major system was added to the chamber, it became the domain of a particular portion of either the collaboration or Fermilab (e.g., the External Muon Identifier, holography, etc.). This type of heterogeneous Fermilab/collaboration social structure is not evidenced in even the largest fixed-target counter experiments at Fermilab, but it is interesting to note that a similar phenomenon does appear with the advent of the mammoth collider detectors like the Collider Detector at Fermilab (CDF), and the D0 detector. I will discuss this further in the final section of the article.

intact. Analysis of other experiments performed at Fermilab reveals copious evidence for experiment strings, and this provides a unifying device for characterizing the entire experimental history of Fermilab.[241] Most importantly, the evidence for experiment strings emerges from the actual practice of scientists and is not an ad hoc taxonomy imposed on the experimental program for the purpose of organizing the sociological and experimental factors of laboratory life.

To this point, the physics and detector driven model has provided a basis for understanding the major continuities that link individual experiments into a string of experiments, but I have not fully described what appears to be one of the deeper motivations for the emergence of such strings. In other words, I have described what the continuities are, but not why they persist through successive iterations of experiments within the context of the resource economies mentioned earlier. As I already indicated, the sociological problems of gaining resources within the economies of proton economics, experimental real estate, and physicist economics can be more difficult to navigate than technical matters because collaborations cannot systematically control them. I will now explain how the case study of experiments E-516, 691, 769, and 791 suggests that experiment strings are one way that physicists attempt to remove many of the physics and sociological uncertainties of performing experiments within the institutional context of Fermilab.

In regard to removing physics uncertainties, using the tagged photon beamline for all four experiments meant that the properties of the beamline transport system were well-known factors, eliminating the need for designing, constructing, and commissioning a new beamline for each experiment.

---

241 For example, there are similar continuities revealed in the E-82, 226, 383, 425, 486, 584, 617, 731, 773 string; the E-531, 653 string; the E-8, 440, 495, 555, 620, 619, 756, 800 string; the E-21A, 262, 320, 356, 616, 770 string; the E-594, 733 string; the E-98, 365, 665 string; the E-1A, 310 string; the E-95, 537, 705, 771 string; the E-70, 288, 494, 605, 608, 772, and 789 string; the E-87, 358, 400, 401, 402, 687, 831 string; and the E-497, 715, 761 string. These experiments display the general characteristics of the physics and detector driven model in the sense that more complex physics measurements demanded a more complex detector design, which gave rise to a more complex social structures for the collaborations. But the opposite was also true. Experiments studying less complex physics measurements had less complex detector designs which resulted in a less complex social structures for the collaborations. My preliminary studies show that similar experiment strings are also found at laboratories like the European laboratory CERN. In regard to photoproduction experiments at CERN during the time periods of E-516 and E-691, there were experiments NA1 (1980 run), NA1 (1983 run), NA14, NA14/2, as well as experiments WA4, WA57, WA58, and WA69.

As indicated earlier, even the design for converting the secondary beam from an electron to a pion beam had been established for almost ten years prior to the proposal for E-769. In addition, using the Tagged Photon Magnetic Spectrometer as the basis for all four experiments made the detector a well-known experimental tool, which could be built upon by the collaboration without facing the experimental problems of understanding the systematic errors inherent in commissioning a new spectrometer.[242] Because the vast majority of the detector subsystems remained intact throughout all four experimental configurations, much of the software that the E-516 collaboration wrote for reconstructing the particle events that occurred in the spectrometer could be reused for the subsequent experiments with simple modifications to the computer code. When the collaboration listed the "job openings" for each new experiment, the major task of writing new software for constantly changing detector configurations and geometries did not have to be included. In the case of E-691, the collaboration's ability to produce physics results shortly after the end of their data run was a direct result of a conscious decision not to modify the majority of the components of the spectrometer from the E-516 configuration.[243] Finally, rather than proposing new, unrelated, physics measurements for each new experiment and increasing the risk of rejection by the Fermilab Directorate, this string chose to focus on a continuous program of studying photoproduction then hadroproduction of charmed particles. The data from each subsequent experiment was an extrapolation that built upon the previous data and forged ahead into new areas of charmed particle measurements, accumulating larger and larger samples, increasingly precise statistics, and more fine-grained measurements.[244]

---

242 In the original E-516 proposal, the collaboration stressed the importance of minimizing the systematic errors of the spectrometer and used this as a justification for viewing the Tagged Photon Magnetic Spectrometer as a "facility" for performing an ongoing series of experiments, rather than simply as a single "experiment." See the *Proposal to Study Photoproduction of Final States of Mass Above 2.5 GeV with a Magnetic Spectrometer in the Tagged Photon Lab*, October 1, 1976, 3.

243 In addition to the evidence of the continuities evidenced in the configuration of the detector subsystems of the two experiments, and the statement of this intent in the written text of the E-691 proposal, I confirmed this in a private communication with R. J. Morrison, a member of both experiments.

244 The success of experiment E-691 is evidenced by the fact that they accumulated 10,000 fully reconstructed photoproduced charmed particle events and made the most precise measurements of charmed particle lifetimes in the world at that time. E-769 accumulated 6,000 fully reconstructed hadronically produced charmed events which was also the largest hadronically produced sample of

But experiment strings also remove many of the sociological uncertainties associated with institutional HEP. The scientific practice of transforming one experiment into another before the first experiment is fully completed helped the collaboration maintain possession of the Tagged Photon Laboratory real estate, thus reducing the probability of being displaced by competing groups. In addition, the fact that the spectrometer was utilized with only modifications to the original E-516 design meant that only the new portions of the experimental configuration were subjected to the most rigorous aspects of the institutional review process prior to approval of the follow-up experiment. While the size and complexity of the Tagged Photon Magnetic Spectrometer powerfully affected the original cost of building the experimental setup, the spectrometer proved to be a good investment of laboratory resources, given the physics results produced.

From the perspective of the collaboration "as a group," the accumulation of expertise for the original "people design" and the fact that this core group participated in the follow-up experiments, allowed them to draw upon these repositories for the expertise needed to carry out the subsequent experiments. From the perspective of the collaboration "as individuals," participating in an ongoing program of successful experiments meant they had a base of operation that assured each physicist continuing future research opportunities.[245]

---

charmed particles produced in the world at that time. Experiment E-791 plans to accumulate 100,000 fully reconstructed hadronically produced charmed particle events during the current running period at Fermilab, which will surpass even the samples accumulated by experiments using the photoproduction mechanism which involves much lower background rates.

245 In a private communication, my colleague Catherine Westfall pointed out that it is important to note why some physicists decide not to continue as part of the core group. She cites two possible reasons for this. On one hand, sometimes the risks involved in joining another experiment are acceptable to experimentalists if they believe it will pay off in terms of physics results that are of major significance. On the other hand, physicists can simply "lose interest" in that program of physics goals. Two examples from E-516 which support her claims are D. Summers and B. Denby who decided to join the CERN-based UA1 collaboration because of the high probability that UA1 would be the first to detect experimental evidence for the (then) elusive W and Z vector bosons. UA1 did in fact discover these particles for which the 1984 Nobel Prize in Physics was awarded to C. Rubbia and S. van der Meer. Upon the completion of his work with UA1, Summers returned as a collaborator on the final two experiments in our case study (E-769, 791), while Denby eventually became a staff scientist at Fermilab and decided to focus his research on what he believed were the more "interesting" physics goals of the Collider Detector at Fermilab collaboration rather than return to this string of experiments (private communication with B. Denby).

In a similar way, the continuity of the institutional affiliations was crucial for the professors and university physics departments that needed access to high-energy particle beams to train graduate students. The importance of the pedagogic function of experiments ought not be underestimated given the resource constraints of physicist economics.[246] In the competitive context of institutional HEP, the longevity of experiment strings can help university-based physicists to publish results more regularly and attract top-notch graduate students to their universities.[247]

But this experiment string also had a subtle but real effect on the Fermilab community's perceptions of the collaboration. With each successive iteration of the spectrometer, the collaboration was viewed by its colleagues as a more competent group that developed a "track record," demonstrating that they could make good on their experimental promises to the Fermilab Directorate.[248] Because the collaboration lowered the risk factor associated with each new experiment, the laboratory not only granted them additional resources more easily, but would also assign them a higher priority than the more "risky" experiments that were still in the design, installation, and commissioning phases. This also had a powerful effect on the beam-management decisions involved in proton economics for each experiment. When the Directorate and the Physics Advisory Committee (PAC) had to decide how to distribute the fixed number of protons from the accelerator, this successful string of experiments, which proposed to extend their studies, was more likely to receive the laboratory's support. The ability to "come through" gives an experiment string an advantage over collaborations that cannot appeal to a long experimental tradition in support of their requests. In fact, at the E-791 proposal presentation to the PAC, the first overhead used by the collaboration

246 Two related motivations for the continuities of experiment strings are the tendency for high-energy physics graduate students to continue to work in a related area of research long after they have received their PhD and the maintaining of professional ties with the university professors who supervised their thesis research.

247 R. J. Morrison of the University of California, Santa Barbara claims that he was still "putting students through" with the very large data sample accumulated with E-691 (private communication). J. Appel of Fermilab claims this is also true of the data taken with experiment E-769 (private communication).

248 The ability to do what they promised was inextricably bound to requests for more and more resources, especially in terms of online and offline computing. Thus, while the collaboration could produce valuable physics results, this heightened the competition with other experiments for laboratory resources.

had a large picture of a slow-moving turtle on it with a caption that referred to the Tagged Photon Magnetic Spectrometer as "Old, Slow, and Successful."[249]

Conclusions and Preliminary Thoughts on Collider Detectors

By describing this case study within the context of the distinct, yet interrelated, resource economies of proton economics, experimental real estate, and physicist economics, I have tried to emphasize that the nature of experimental practice in high-energy physics must be studied within the larger social context of laboratory life. By abstaining from the temptation to describe experiments as closed systems, which can be divorced from the laboratory context, I have tried to capture some of the most salient aspects of the actual practice of high-energy physics. Using a physics and detector driven model for social structure, I attempted to show how experimental HEP is an interrelated "human-factors-like" system consisting of physics measurements, spectrometer designs, and people designs working together as a unit. The physics and detector driven model also explains why the social factors, which are noticeably absent in scientists' accounts of science, cannot be ignored if the goal is to describe what physicists actually *do* in experimental halls. By using a case study that is representative of many HEP experiments within the time period of 1976-1991, I have tried to offer an alternative to the image of high-energy physics research being dominated by a continuing series of crucial discovery experiments. While the data produced by E-516, 691, 769, and 791 did not substantially influence the overall conceptual development of HEP, I claim that the account presented here is very typical of what most scientists actually do at Fermilab.

---

249 S. Bracker who was a member of E-516, 691, 769, and 791 claims that the attributes of "Old, Slow, and Successful" were coined by theorist J. D. Bjorken who at that time was Fermilab's Associate Director for Physics Research. During the era of Fermilab's first Director, Robert Wilson, there were multiple examples where although experimental risk factors were high, the laboratory still assigned the proposal a high-level of priority because of the promise of important and unique physics results (for example, experiment E-1A). But the collaborations which proposed these more risky experiments had to remove as many of the physics and sociological uncertainties as possible. As pointed out in Galison's account of E-1A, this collaboration failed to do this. During the post E-516 era of Fermilab's second Director, Leon Lederman, there was a decided shift toward larger and more complex experimental "facilities" as part of the Tevatron II Project which upgraded the fixed-target experimental areas to 800 GeV capabilities. With the development of increasingly large, complex, and costly detectors, substantial experimental risks became less acceptable to both collaborations and the Fermilab management.

I have also argued that the physics and detector driven model can be used to describe strings of experiments that are motivated partially by experimentalists' attempts to avoid many of the uncertainties of proton economics, experimental real estate, and physicist economics. Experimentalists must learn to trade with the commodities of protons, experimental halls, and physics expertise because currently there is no other way to participate in accelerator-based high-energy physics research. I attempted to support my claim that the formation of experiment strings is an established scientific practice by demonstrating how copiously they appear throughout Fermilab's entire research program. Moreover, using experiment strings as a unifying device, which emerges from the actual practice of scientists shows how crucial the experimentalists' role is in writing a veridical account of scientific practice; that is, if the goal of such accounts is to describe what high-energy physicists actually do in the laboratory.

Preliminary research shows that the physics and detector driven model and the existence of experiment strings in fixed-target counter experiments may also be suggestive about the sociological and experimental factors associated with large Fermilab collider experiments like the Collider Detector at Fermilab (CDF) and the D0 collider detector.[250] Within the constraints of the resource economies mentioned previously, the model would predict that more complex physics measurements would demand a corresponding increase in the size, cost, and complexity of the detector designs, which would give rise to more complex social structures for these collaborations. One would begin such a study with a diagram of the overall detector and a list of the institutions responsible to design and construct the detector subsystems. If the claims presented thus far are correct and if they can be extended to collider detector collaborations, this exercise should define the basic social structure of the more than 250 physicists in the CDF collaboration. Analysis of the early phases of CDF (design and construction) shows that the expertise of the collaboration was in fact distributed along the lines drawn in a diagram of the CDF detector (track chambers, calorimeters, muon systems, and electronics), with the institutional responsibilities for the detector subsystems remaining the repositories of expertise throughout all phases of the experiments performed with the CDF detector.[251]

---

250 Fermilab's accelerators can be used to produce beams of protons and antiprotons which are circulated then collided together in the center of large collider detectors like CDF and D0.

251 For the details of which institutions assumed responsibilities for the CDF detector subsystems

This research also shows that the complexity of the physics measurements to be made drove the size, cost, and complexity of the CDF detector to a level that far exceeded the experiments in our case study ($60 million, 2,000 tons, 100,000 channels of electronics).[252] One would subsequently expect that a far more complex higher-level social structure would be superimposed on the basic social structure by the laboratory in order to manage the detector facility and the collaboration. The history of the CDF collaboration shows that this was evidenced in two major ways. First, Fermilab insisted that CDF's higher-level management structure be headed by co-spokespersons, one physicist from a university in the collaboration and one physicist who was a Fermilab scientist. With a detector of this size, complexity, and cost, the risks that might otherwise be taken with moderately sized fixed-target counter detectors became absolutely unacceptable to the Fermilab management and the experimentalists. This model of joint spokespersons was one way that the laboratory could maintain control over all aspects of the CDF collaboration's activities.

A second way that Fermilab maintained control over CDF was by forming a Fermilab-based CDF department devoted solely to the operation, maintenance, and upgrading of the detector. With the formation of the CDF department, we see the emergence of an even more complex and heterogeneous social structure that adds new dimensionality to the physics and detector driven model described thus far. Because even the largest and most complex fixed-target counter experiments at Fermilab did not have dedicated support departments, one could initially conclude that this represents a novel sociological development in HEP. But closer examination shows that the formation of the CDF department has organizational links to the dedicated support departments that operated and maintained large bubble chamber facilities like the Fifteen Foot Bubble Chamber at Fermilab.[253] While the existence of a laboratory/university social structure is evidenced in both cases, the formation of these support departments appears to be motivated by different issues. The emergence of dedicated support

see *The Collide, Detector at Fermilab; A Compilation of Articles Reprinted from Nuclear Instruments and Methods in Physics Research-A* (Amsterdam: North-Holland Physics Publishing, 1988).

252 For the technical details of the *CDF detector see the Design Report for the Fermilab Collider Detector Facility* (Batavia, IL: Fermilab, 1981).

253 For a description of the history of the operation of the Fifteen Foot Bubble Chamber, see M. Bodnarczuk (ed), *Reflections on the Fifteen Foot Bubble Chamber at Fermilab* (Batavia, IL: Fermilab, 1989).

departments for large bubble chambers was motivated (primarily) by the safety problems associated with using liquid hydrogen in the chamber and (secondarily) by the size and complexity of the chamber.[254] The formation of the CDF department appears to have been driven primarily by the size, complexity, and cost of the detector, not by safety-related issues.

The previous discussion of experiment strings would also predict evidence for continuities between multiple experiments performed with the CDF detector; the physics goals, the detector design, and the ability of the core group of physicists to work together as a unit in a continuing program of successful physics studies. To date, four experiments have been proposed using the CDF detector (E-741, 775, 775A, and 775B), all of which display many of the properties of an experiment string.[255] Analysis of the respective experimental proposals shows that each experiment was a successive iteration of upgrades to the basic detector subsystems and associated computing systems.[256]

But the physics and detector driven model and the notion of experiment strings may also be suggestive about the next generation of collider detector collaborations like the Solenoidal Detector Collaboration (SDC) at the proposed Superconducting Super Collider (SSC) Laboratory. If fully approved,

254 Galison points out how important safety issues were when L. Alvarez decided to use liquid hydrogen in his seventy-two-inch bubble chamber. Given the volume of liquid hydrogen need for Fermilab's Fifteen Foot Bubble Chamber, these safety related problems became enormous. See Peter Galison, "Bubbles, Sparks, and the Postwar Laboratory," in L. M. Brown, M. Dresden, and L. Hoddeson (eds.) *Pions to Quarks; Particle Physics in the 1950s* (New York: Cambridge University Press, 1989), 220.

255 Experiment E-741 was approved by the Fermilab Directorate in April 1982; E-775 was proposed in May 1986 and approved in July 1986; E-775A was proposed in June 1988 and approved in January 1989; E-775B was proposed in September 1988 and received Stage I approval from the laboratory in January 1989.

256 Another area of research that is beyond the scope of this article involves instances where experiment strings seem to be transmitted from fixed-target counter to large collider detector experiments. Gordon Fraser at CERN pointed out that a major factor in the formation of the core group for the Aleph collider detector at CERN's Large Electron Positron (LEP) Collider was the admiration and loyalty that experimentalists had for spokesperson, Jack Steinberger. In this case, Steinberger himself became a major continuity around which the core group for Aleph was built. Although there were almost no continuities between the physics interests/detector designs of the fixed-target neutrino counter experiments at CERN's Proton Synchrotron in the 1960s and this LEP experiment, Fraser claims that Steinberger and a core group of physicists from these neutrino experiments were the continuity around which the larger Aleph collaboration was formed. See G. Fraser, "Aleph" in *CERN Courier*, vol. 31, no. 1, January/February 1991, 1-4.

SDC will begin running at the SSC Laboratory around the year 2000. As the model would predict, the complexities of the proposed physics measurements to be made have demanded that its design be among the largest, most costly, and complex devices ever designed for any scientific activity.[257] This would suggest that an extremely complex "people design" for the collaboration would have to emerge to carry the experiment through to completion and that it would be based on the basic structure of the detector subsystems—and so it is. The basic social structure of the SDC collaboration is composed of nine Technical Steering Committees (The Calorimetry, Muon Systems, Superconducting Magnet, Tracking, Computing and Analysis Software, Electronics and Data Acquisition, Detector Integration and Experimental Facilities, and Physics and Detector Performance Committees), each headed by a chairperson from the collaboration. In other words, the SDC collaboration's organization chart is a sociological representation of the list of detector subsystems that compose the experimental apparatus. The physics and detector driven model would also predict an increase in the complexity of the higher-level social structure needed to manage the SDC collaboration. Currently, there are about 600 physicists in the collaboration, with a higher-level social structure consisting of one Spokesperson, three Deputy Spokespersons, a Technical Manager, and a 13 member Executive Board.[258] The SDC collaboration has written and ratified a Governance Document which describes the organization and management of the group and the rules for admitting additional members into the experiment.[259]

It is interesting to speculate about the present trend toward the use of collider detectors and away from fixed-target counter detectors. Currently, fixed-target counter experiments are beginning to play less of a role in the

257 Current estimates by the SDC collaboration describe a detector that will cost $500 million (in 1990 dollars), weigh 10,000 tons, and have 20 million channels of electronics. For details, see the *Solenoidal Detector Collaboration Expression of Interest*, 24 May 1990.

258 The higher-level social structure of CERN's Delphi experiment at LEP, has been referred to as a "parliament." The collaboration consists of over 500 physicists, and is headed by a Collaboration Board which is responsible to reconcile potential conflicts of interest between participating institutions and overview the final production of physics results. In addition, a collaboration Executive Committee is responsible to make technical recommendations for the experiment which must be subsequently ratified by the Collaboration Board. See Gordon Fraser, "Delphi," in *CERN Courier*, vol. 30, no. 8, November 1990, 1-5.

259 For details, see the *Solenoidal Detector Collaboration Expression of Interest*, 24 May 1990, 93 ff.

overall research of HEP. If this trend continues into the SSC era, the social structure of US HEP might be defined by the number of interaction regions in the world-wide network of accelerators, the distribution of the majority of high-energy physicists into the collaborations that propose collider detectors, and a social structure for these collaborations based upon the model I have described above.[260] This type of localization of the majority of high-energy physicists into a small number of large collider experiments, might even suggest that the size, cost, and complexity of these detectors may become a self-limiting factor in the sociological makeup of US HEP.

In regard to the model's prediction that such detectors would be used to perform strings of experiments, even in the early design phase of the SDC, the collaboration is trying to anticipate the need for upgrading the capabilities of the detector to do subsequent experiments. This is partially due to the SSC laboratory's proposed plan to eventually upgrade the accelerator's ability to produce particle interactions (luminosity) and the collaboration's plan to pursue a continuing program of physics goals that spans multiple experiments.[261] In addition, when detectors become as large, complex, and costly as the SDC, the importance of removing the uncertainties mentioned in our case study is crucial. With SDC, the failure to eliminate as many of the risks as possible is absolutely unacceptable to the experimentalists and the SSC laboratory management, because the failure of the detector to perform as intended could

260 Currently, one Expression of Interest has been submitted to the SSC Laboratory for a 20 TeV Super Fixed Target Beauty Spectrometer (SFT) which is estimated to cost about $24 million (in 1990 dollars). Because the physics measurements to be made with the SFT are far more complex than those described in our case study, the spectrometer design exhibits a corresponding increase in cost, complexity, and size. The various components of the SFT include two large analysis magnets, a silicon microvertex detector, several stations of PWC wire and pad chambers, a Ring Imaging Cerenkov Counter, and electromagnetic detector, and a muon detector. These detector subsystems are distributed over a spectrometer which is 70 meters long and 10 meters wide at the downstream end. Based on my current analysis, one would also expect that the complexity of the social structure of this collaboration would also increase substantially. Currently, there are about 140 physicists affiliated with this experiment, and even in this early design phase, they are organizing themselves around the lines drawn in the SFT spectrometer diagram. See *An Expression of Interest in a Super Fixed Target Beauty Facility (SFT) at the Superconducting Super Collider*, May 25, 1990.

261 Luminosity is a measure of how effectively a collider induces collisions between the two circulating beams of particles. It is the product of the number of particles per second circulating in one beam and the average number of intercepted particles per unit transverse area in the other beam. When multiplied by the cross section for a given reaction, it gives the average number of events per second for that reaction. For a brief account of the physics goals of SDC and the technical details of the anticipated luminosity upgrade, see D. Green, *SDC at High Luminosity, Fermilab-Conf-90/110.*

adversely affect a major portion of the US high-energy physics community for a decade or more.[262] While it may not be clear exactly how a collaboration like SDC will function throughout the lifetime of the detector, the notions of proton economics, experimental real estate, physicist economics, the physics and detector driven model of social structure, and experiment strings together may provide a useful way to characterize the nature of institutionalized scientific practice in high-energy physics into the next millennium.

262 Current estimates claim that approximately 50 percent of the HEP community in the U.S. will be involved in some aspect of the two collider experiments at the SSC Laboratory, with 50 percent participating in research at other U.S. laboratories like Fermilab and SLAC. For details see the *HEPAP Subpanel on the U.S. High Energy Physics Research Program for the 1990s*, U.S. Department of Energy Office of Energy Research Division of High Energy Physics, DOE/ER-0453P, April 1990, 68 ff.

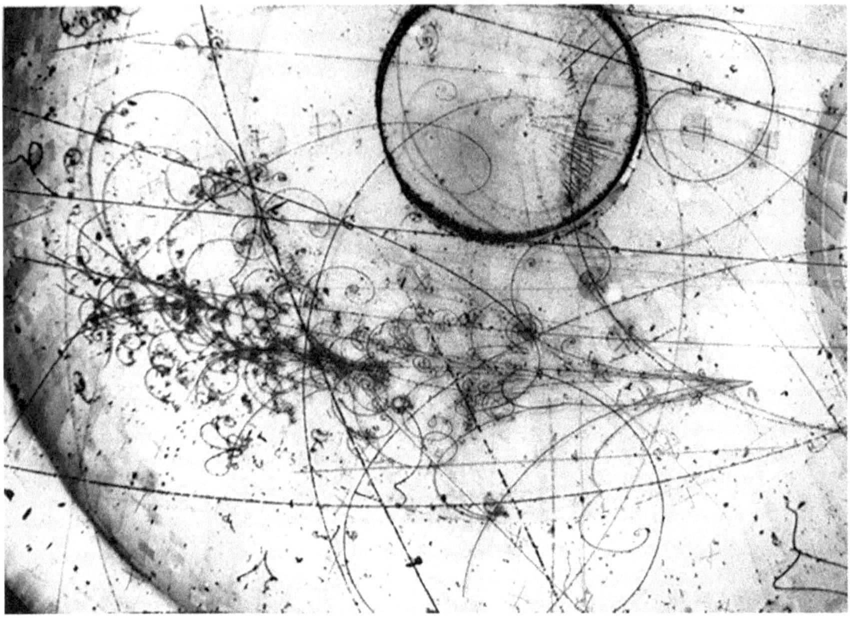

Picture of neutrino interaction in the Fermilab 15-Foot Bubble Chamber with heavy neon-hydrogen liquid mixture taken in April 1976. Nearly one neutrino interaction per picture was found with that run targeting 1013 protons at 400 GeV with the wide band, two-horn system. Frequently the chamber was flooded with tracks from several neutrino interactions in the same exposure. In addition to increasing the interaction rate, the heavy neon mixture allowed many of the particles from neutrino interactions to be recognized by direct inspection of the track appearance: protons, charged pions, and kaons produced secondary interactions; neutral pions were evidenced by their gamma rays converting to electron pairs; muons sailed right through the liquid without interacting and direct electrons or positrons from the vertex were recognized by successive kinks and associated gamma ray conversions along their tracks. A major interest in this experiment by a Columbia University-Brookhaven Laboratory collaboration was the study of "di-lepton" events in which two muons or a muon and an electron were produced in high energy neutrino interactions (See "Picture of Neutrino Interaction" in the Fermilab *The Village Crier* Vol. 8 No. 18, May 6, 1976, p 1.)

5

Reflections on the Fifteen Foot Bubble Chamber at Fermilab[263]

One of the most difficult aspects of high-energy physics is the pace at which change occurs as the discipline is advanced by theory, experiment, and technological breakthroughs. Perhaps one of the best examples of this is the discovery of the transistor followed closely by the advent of the digital computer. Today, if a data acquisition system takes more than 1–2 years from the design phase to implementation, not only will the components probably be available commercially in off-the-shelf products, but the technology may even be out of date. Even some theories seem to come and go almost faddishly. But one aspect of high-energy physics that tends to change more slowly is the general types of detectors used in experiments. In fact, one gets an interesting view of the development of particle physics by analyzing the trends in the types of detectors that have been used and how older-style detectors have been retired as more sophisticated devices, using new technologies, have been used to revise and reconfigure experiments.

Over the last 25 years, high-energy physics detectors have been classified in a number of different ways, with two of the most widely used categories being electronic counter and bubble chambers apparatus. Ever since Donald A. Glaser developed the first bubble chamber in 1952 at the University of Michigan, bubble chambers have been used successfully in myriad high-energy physics experiments including the discovery of the ρ, η, Ξ^0 and Ω^- particles. Glaser's original bubble chamber was a transparent glass device which was only a few inches across. He knew that if he placed a pure liquid in a sealed container, he could superheat

263 This was originally published as the introduction to, Mark Bodnarczuk (ed), *Reflections on the Fifteen Foot Bubble Chamber at Fermilab* (Batavia, IL: Fermi National Accelerator Laboratory, 1989).

the liquid beyond its normal boiling point, and the liquid would not boil. Once superheated, boiling could be triggered by something penetrating into the liquid in the vessel. Glaser calculated that the energy that charged particles would deposit in the superheated liquid while penetrating and ionizing the chamber would be enough to trigger a trail of boiling bubbles along the trajectory of the particle. He further surmised that these bubble events could be observed with the naked eye through the glass chamber or permanently recorded by taking a photograph of the interaction in the chamber before the bubbles dispersed. Glaser later won the Nobel Prize for this innovative work.

Shortly after this first design proved successful, bubble chambers were constructed from metal-glass combinations, and the photographing was done via optical glass windows built into the chamber wall. Eventually, a group at Berkeley, headed by Luis Alvarez, extrapolated this technology, moving from a chamber that was only a few inches across to one that was six feet long. Although an extension of Glaser's idea, Alvarez's chamber involved many new innovations, some of which permanently changed the field of particle physics. As well as providing a clear advance in physics potential due to the much larger fiducial volume in which events could occur, the large bubble chamber was a clear and prophetic departure from the kind of "table-top" experiments that Glaser's device typified.

Whereas Glaser was directly involved in all phases of the design, construction, and operation of his bubble chamber (in the tradition of experimenters like Boyle, Torricelli, or Newton), Alvarez assembled a team composed of structural, cryogenic, and operations engineers plus other technical personnel in order to carry his experiments through to completion. This was complemented by physicists and film scanners to process the data once it was accumulated. Alvarez's collaborative efforts and the idea of assembling an experimental support team, many of whom were not physicists, was a prototype of all subsequent large bubble chamber runs at high-energy physics laboratories like Fermilab, Brookhaven, SLAC, and CERN. Even more interesting is the prototypic role that these experiments have played in describing the type of support structure necessary to carry out today's large electronic counter experiments. Alvarez's bubble chamber changed the way particle physicists designed and configured their experiments, establishing an outstanding potential for high-energy physics research at Lawrence Berkeley Laboratory (LBL) and Stanford Linear Accelerator Center. Alvarez won the Nobel Prize for this work.

Fermilab's 15-Foot Bubble Chamber is the last in a long developmental line of large bubble chambers in the United States and in Europe. The following are only a few highlights of this illustrious history. On the Continent, CERN began work on the Two Meter Hydrogen Bubble Chamber in 1959. The two-meter chamber ran successfully at both the Proton Synchrotron and the Super Proton Synchrotron. The Gargamelle Bubble Chamber was installed at CERN in 1969 and was used to probe the nucleus by inelastic scattering with a neutrino beam, providing evidence for quarks as the constituents of nucleons, and later, conclusive evidence for the existence of neutral currents. Other big bubble chambers across the Atlantic included CERN's Big European Bubble Chamber (BEBC), and the Mirabelle Bubble Chamber designed at Saclay, France, for use at the 70 GeV accelerator at Serpukhov, USSR.

In the United States, Tom Fields and a group from Northwestern University developed the first bubble chamber in a superconducting magnet at Argonne National Laboratory (ANL), leading the way for Gale Pewitt and his colleagues who developed the 12-foot bubble chamber, also at ANL. The seven-foot bubble chamber at Brookhaven (BNL) also produced interesting physics and new discoveries for a considerable period of time. In this tradition, Fermilab's 15-Foot Bubble Chamber is the last large cryogenic bubble chamber to operate anywhere in the world and its decommissioning marks the end of an era. The first tracks were recorded on September 29, 1973, in conjunction with experiment E-28, and the last tracks were recorded on February 1, 1988, in conjunction with experiment E-632. In all, there were 17 approved and completed experiments performed with the chamber, covering a wide variety of physics topics.

There were also two experiments carried out with the 15-Foot Bubble Chamber Magnet, using cosmic rays rather than beams from the accelerator. Charles Baltay (Yale University) described some of the unique aspects of the 15-Foot Bubble Chamber, namely the ability to see electrons and strange particles (K^O and Λ^O) and the ability to detect hadrons and the fine details of decay vertices.[264] The chamber provided unique contributions to neutrino physics, including charm production by neutrinos, rare neutral current processes, and tests of QCD through hadronic production by neutrinos.[265]

---

264 See Mark Bodnarczuk (ed), *Reflections on the Fifteen Foot Bubble Chamber at Fermilab* (Batavia, IL: Fermi National Accelerator Laboratory, 1989), 3.

265 For a complete list of all experiments performed with the 15-Foot Bubble Chamber and a list of all physics publications to date that have been based upon data taken with the 15-Foot Bubble Chamber see Mark Bodnarczuk (ed), *Reflections on the Fifteen Foot Bubble Chamber at Fermilab* (Batavia, IL: Fermi National Accelerator Laboratory, 1989), Appendix A.

Stephen Hawking, Lucasian Professor of Mathematics at Cambridge University, and author of the best-selling book, *A Brief History of Time*, arrived at Fermilab on May 10, 1989, to participate in a Wormhole Workshop organized by the Fermilab Astrophysics Group. Pictured above, Rocky Kolb, co-head of the Fermilab Astrophysics Group, introduced Professor Hawking at a press conference. Afterward, Hawking gave a public lecture entitled "Black Holes and Their Children, Baby Universes." During his lecture, Hawking said that the two most important cosmological questions to be answered are to discover why the cosmological constant is zero and why space-time has three dimensions of space and one of time. "Wormholes can probably answer the first question and possibly the second," he concluded. Following his public lecture, Hawking was asked what may have existed before the Big Bang. He responded, "You can't go back beyond the Big Bang. Time is defined only within the universe. To ask what happened before the Big Bang is like asking what happens north of the North Pole."

6

Science as Knowledge, Practice, and Map Making[266]

In a world that's been powerfully shaped and defined by modern science, it's important to reflect on its historical and philosophical roots to understand the *purpose* of science. The graveyard of the philosophy of science is littered with bodies of knowledge that are interesting but widely discarded in this regard. I'm thinking specifically about the views of Rudolph Carnap's logical positivism, with its focus on meaningful scientific discourse, and scientific propositions being verifiable, and Karl Popper's view that scientific theories can never be verified, but rather a proposition is "scientific" only when it can be falsified. Both tried to press science into the service of being the ultimate adjudicator of the veracity and meaning of all human knowledge. Neither one had ever worked in a science laboratory.

Carnap's views echoed an empirical, inductive, bottom-up, Baconian perspective where scientists verified scientific theories by making observations and gathering them into increasingly generalized scientific propositions that could be verified. His vision was to unify human knowledge using this methodology—something he began to concretize in a multiple volume series entitled, *The Encyclopedia of Unified Science*.[267] The foundation stone of Carnap's logical positivism was the proposition, "Only that which can be verified has

---

266 This paper was originally published as, *Using Research Metrics to Evaluate the International Atomic Energy Agency Guidelines on Quality Assurance for R&D*, in June of 1994 when I was in the director's office of the National Renewable Energy Laboratory in Golden, Colorado, see NREL/TP-320-6696, UC Category: 900, DE94011823.

267 Otto Neurath, Rudolf Carnap, and Charles W, Morris (eds), *International Encyclopedia of Unified Science*, Volume I, Nos. 1-5 (Chicago: The University of Chicago Press, 1955).

meaning."[268] But as Popper pointed out, this statement itself cannot be verified, so by definition, it has no meaning.

Popper's views echoed a theoretical, deductive, top-down, form of Humean skepticism that rejected inductive reasoning as a method for verifying scientific theories.[269] Commonly called, Hume's Problem of Induction, Hume argued against the assumption that the future would resemble the past, so for Popper, scientific predictions about yet unobserved phenomena could not be verified based on previous observations, regardless how big the sample of previous observations was.

For example, one could not say the following with any certainty, "Every swan that Europeans have seen for the last 1,000 years has been white; therefore, all swans are white." Popper argued that there was always the possibility that we might someday encounter a black one, which would falsify the theoretical prediction. So, while a hypothesis that has passed myriad tests could be *corroborated* by more and more data points (i.e., more swan sightings), it could never ultimately be verified, only falsified.

Decades before Carnap and Popper developed their perspectives on science, Friedrich Nietzsche's groundbreaking book, *Twilight of the Idols, or, How to Philosophize with a Hammer* (published in 1889), diagnosed the tendency of later philosophers of science to see science as severed from history, i.e., the actual practice of science. Nietzsche argued that we should no longer show undo "respect" for disciplines like philosophy of science and science itself by dehistoricizing them.[270] Even today, many philosophers of science and practicing scientists tend to view science as being about producing scientific knowledge (outcome), and they tend to ignore the actual practice of science and the day-to-day realities of laboratory life (process).

But the reality is much more complicated than this. The advent of quantum mechanics in the mid-1920s by scientists like Niels Bohr, Erwin Schrodinger,

---

268 See Rudolph Carnap, "The Elimination of Metaphysics Through Logical Analysis of Language" in A. J. Ayer (ed), *Logical Positivism* (Los Angeles, CA: Free Press, 1959), Rudolph Carnap, *Testability and Meaning* (Whitefish, MT: Literary Licensing LLC, 2011), and Rudolph Carnap, *Philosophy and Logical Syntax* (Los Angeles, CA: Free Press, 1935).

269 See Karl Popper, *The Logic of Scientific Discovery*, 2nd Edition (New York: Routledge Classics, 2002), and David Hume, *An Enquiry Concerning Human Understanding* (Oxford, UK: Oxford University Press, 2008).

270 Friedrich Nietzsche, *Twilight of the Idols* (New York: Hackett Publishing Company, Inc., 1997), chapter 2.

Werner Heisenberg, Max Born, Paul Dirac,[271] plus the 1964 publication of Thomas Kuhn's, *Structure of Scientific Revolutions*,[272] Berger and Luckmann's, *The Social Construction of Reality*,[273] Ian Hacking's, *Representing and Intervening*,[274] and Peter Galison's, *How Experiments End*,[275] reshaped the discussion in the history and philosophy of science to dialogue between postmodernist, deconstructivist views of science (subjective, individual), and objective realist view science (objective, physical). I will discuss these issues in a subsequent article. For now, suffice to say that the constructivist-realist debate made it even more difficult to press science into the service of being the ultimate adjudicator of the veracity and meaning of all human knowledge.

Science is not, and cannot be, the only yardstick against which everything else is measured. Despite the ambition of high-energy physicists to develop a "theory of everything" using the Standard Model, Peter Galison argues, "Recent years have seen a turn largely against the rhetoric of unity [of science], ranging from the pleas of condensed matter physicists for disciplinary autonomy all the way to discussions in the humanities and social sciences that involve local history, feminism, multiculturalism, postmodernism, scientific relativism and realism, and social constructivism."[276] Being the ultimate adjudicator of truth and meaning is not the purpose of science. But unfortunately, like Abraham Maslow taught us, "He who is good with a hammer thinks everything is a nail."[277]

---

271 See A. P. French and P. J. Kennedy (eds), *Niels Bohr: A Centenary Volume* (Cambridge, MA: Harvard University Press, 1985), P. A. M. Dirac, *The Principles of Quantum Mechanics*, 4th ed. (Clarendon Press, 1982), Max Jammer, *The Conceptual Development of Quantum Mechanics* (New York: McGraw-Hill, 1966), and Max Jammer, *The Philosophy of Quantum Mechanics: The Interpretations of Quantum Mechanics in Historical Perspective* (New York: Wiley, 1974).

272 See Thomas Kuhn, *The Structure of Scientific Revolutions: 50th Anniversary Edition*, 4th Edition (Chicago: University of Chicago Press, 2012).

273 See Peter Berger and Thomas Luckmann, *The Social Construction of Reality: A Treatise in the Sociology of Knowledge* (New York: Anchor Books, 1967).

274 See Ian Hacking, *Representing, and Intervening: Introductory Topics in the Philosophy of Science* (New York: Cambridge University Press, 1983).

275 See Peter Galison, *How Experiments End* (Chicago: The University of Chicago Press, 1987).

276 See Peter Galison and David Stump (eds), *The Disunity of Science: Boundaries, Contexts, and Power* (Menlo Park, CA: Stanford University Press, 1996).

277 See Abraham Maslow, Psychology of Science: A Reconnaissance (Washington, DC: Gateway Editions, 1969), 15.

But while science should not and cannot stand alone as the final judge of what's true and meaningful, it is a powerful benchmark against which all other knowledge and experience can be and should be evaluated and calibrated. Science is necessary, but not sufficient to bear that burden of being the sole criterion of determining truth and meaning, without other criteria that may, or may not be reducible to law-like, cause and effect verification or falsification.

So, what is the purpose of science? The purpose of science is to *map out* the visible and invisible, inanimate and animate constituents, forces, and processes of a universe that came into existence about 13.8 billion years ago in the Big Bang, and to harness them to serve the purposes of human beings. I refer to this as the cartographic function of science—the map-making function.[278] In the same way that early explorers had to physically travel through an unexplored region of the earth (drawing maps of that region as they go), experimental scientists use their experimental apparatus as extensions of their physical senses to explore uncharted physical domains, drawing experimental maps of the measurements and newly created effects and phenomena they encounter. If the experimental map (i.e., scientific publications) is drawn accurately, it enables other experimenters to move directly to the frontier of scientific knowledge and continue mapping out physical effects that lay beyond that point.

Because most experimental results obtained at places like Fermilab or the SLAC National Accelerator Laboratory at Stanford University are published as "pre-prints" and presented orally at conferences long before they appear in a professional journal, primary publications in journals have much less to do with the transmission of scientific knowledge and much more to do with establishing priorities for scientific discoveries. Published results become valuable commodities that are traded back to the laboratory or funding agency in exchange for more resources to perform more experiments.[279] It is the practice

278 Thomas Kuhn originally used the metaphor of a map to describe how scientific paradigms shape and define the research programs of scientists. In the case of high-energy physics, the paradigm includes the theoretical and experimental evidence that is the basis of the Standard Model, plus the types of experimental methods, detectors, data analysis, and practices described in this book. See Thomas Kuhn, *The Structure of Scientific Revolutions*, 2nd edition (Chicago, IL: The University of Chicago Press, 1970), 109.

279 In a case study of high-energy physics experiments performed at Fermilab, I've described a number of distinct, yet interrelated, resource economies, (each having its own commodity) within which high-energy physics collaborations must trade to obtain the resources needed to perform experiments. The resource economies and commodities are proton economics (high-energy protons from the accelerator), experimental real estate (a place to setup the apparatus), physicist economics

of science that most appropriately defines what basic experimental science is, not the experimental maps that are subsequently published in journals and traded back to funding agencies for additional resources.

Part of the confusion about science-as-knowledge (outcome) and science-as-practice (process) is the heavy semantic burden that the word "science" and its derivatives have been forced to bear. While the word science does come from the Latin *Scientia* (knowledge), its primary meaning with reference to experimental science following the 17th century scientific revolution is the experience of vexing nature to understand its properties.[280] For Galileo, the act of doing an experiment (not the voyeuristic study of Aristotelian "texts") produced a kind of knowledge that could not be obtained in any other way.[281]

Echoing Nietzsche's admonition about the failure to ground knowledge in history, published scientific papers (science-as-knowledge) are produced after-the-fact, downstream of the creation of physical effects in the experimental apparatus (science-as-practice), many of which do not occur in nature unattenuated.[282] So, it's the planning, managing, operating, documenting, and recording activities that go into the production of a scientific publication (science-as-practice), that create the outcome (science-as-knowledge), and together they constitute the map-making function and purpose of science.

A more general sociocultural framework within which to describe the purpose of science is the essential tension between the Baconian and Cartesian values about science. On one hand, those holding to Baconian values tend to view science as the servant of the taxpayers and shareholders

(physics expertise), computing economics (online and offline computational power), and physics economics (scientific publications that are traded back to the Laboratory Director for more resources). See Mark Bodnarczuk, *The Social Structure of Experimental Strings at Fermilab; A Physics and Detector Driven Model,* Fermilab-Pub-91/63, March 1990, 2 ff.

280 I am thinking primarily of Francis Bacon, see Francis Bacon, James Spedding, Robert Leslie Ellis, Douglas Denon Heath (eds.) *The Works of Francis Bacon* (Boston: Taggard and Thompson, 1893); and Robert Boyle, see Robert Boyle, *The Works of the Honorable Robert Boyle* (London, 1772).

281 See Galileo Galilei, *Two New Sciences - Including Center of Gravity and Force of Percussion,* trans. by Stillman Drake (Madison, WI: The University of Wisconsin Press, 1974), 66.

282 Hacking claims that certain effects (like the Hall effect) do not exist in nature outside of certain kinds of apparatus. While passing a current through a conductor at right angles to a magnetic field will produce a potential anywhere in nature, nowhere outside the laboratory (the technological devices we construct), is there such a pure arrangement. See Ian Hacking, *Representing and Intervening* (New York: Cambridge University Press, 1987), 225 ff.

who finance it. The knowledge produced should be utilitarian, i.e., applied science that improves the quality of life through technological advances.[283] Alvin Weinberg's model of internal and external criteria for evaluating which types of science should be funded exemplifies Baconian values. His internal criteria ask two questions. First, is the field ready for exploitation? Second, are the scientists in the field actually competent? Weinberg claimed that these decisions could be made only by scientists. He identified three external criteria that could be decided by nonscientists: technological merit, social merit, and scientific merit. The criterion of scientific merit assessed the degree to which the knowledge produced by the discipline requesting funding contributed to its neighboring disciplines.[284]

On the other hand, those holding to Cartesian values tend to view basic experimental science as the "feedstock" of future utilitarian (applied) science. They argue that all cultured societies should support scientific knowledge for its own sake (like art or music), even if there are no currently known applications of the knowledge produced. While these two views seem orthogonal, there is actually a dialectic tension between Baconian and Cartesian values, and the outcome of this dialogue is crucial to developing a science policy. Within this macrosocial framework, there are important microsocial aspects of the purpose of science that appear in the essential tension between science-as-knowledge and science-as-practice, as evidenced in individual scientific research proposals at basic research laboratories that I've worked at like Fermilab and SLAC National Acceleratory Laboratory at Stanford University.

When scientists propose research programs at one of the DOE-funded National Laboratories they are required to concretize them into an experimental proposal, and these proposals become an important part of the information that the laboratory uses to decide whether to conduct the research.[285] In these experimental proposals, scientists make two types of claims. First, they make knowledge claims that are calculated hypotheses of the experimental path that they want to pursue, and the expected results and new knowledge that will be

283 Francis Bacon, *The Works of Francis Bacon,* James Spedding, Robert Leslie Ellis, Douglas Denon Heath (eds.) (Boston: Taggard and Thompson, 1893).

284 See Alvin Weinberg, "Criteria for Scientific Choice," in *Minerva,* vol. 1, 1963, 159-171.

285 A description of the DOE-funded National Laboratories in the Office of Science can be found here: https://www.energy.gov/national-laboratories.

obtained by performing the experiment, i.e., the territory they plan to map out. Second, they make practice claims that define the planning and managing elements needed to obtain that new knowledge, including estimates of how long it will take to travel the experimental path, the types of experimental equipment that will be needed, how much online and offline computing resources will be needed, how much engineering and technician support will be needed, and other costs and resources.

Experimental proposals are often submitted two or more years prior to performing the experimental work, and because of the exploratory nature of science the knowledge claims are like moving targets that may not be well defined. Consequently, the practice claims needed to obtain that knowledge are often "placeholders" that cannot be nailed down. But the knowledge claims become increasingly refined and fine-tuned as they move toward the performance of the current year's research, and, as a result, scientists try to articulate and evaluate the practice claims more precisely. But nailing down practice claims (i.e., managing the research) has been problematic for reasons that I will explore below.

Metrics for Science-as-Knowledge

I will use the notion of a "metric" to indicate a standard of measurement. But I want to make the distinction between a direct metric (a close logical, causal, or consequential relationship to the things being measured) and an indirect metric (a collateral or circumstantial relationship to what is being measured). Direct metrics are preferred because they provide a more veridical characterization of the physical phenomena. So, when evaluating whether an experiment has produced "good" physics, one could use Irvine and Martin's indirect metric of counting the number of citations that a publication receives. But this assumes that if a scientific paper contains "good" physics, it will be cited more frequently than one that does not.[286]

One could also use the number of Nobel Prizes received by Americans as a metric of the effectiveness of US science policy or increases in a laboratory's growth in funding and staffing levels could be used as an indicator of the quality of its research. But increases in funding or staffing levels may be due

286 See J. Irvine and B. R. Martin, "Basic Research in the East and West: A Comparison of the Scientific Performance of High-Energy Physics Accelerators," in *Social Studies of Science,* 1985, vol. 15, 300, and Steven Yearly, *Science, Technology, & Social Change (*London: Unwin Hyman, 1988), 88 ff.

largely to political factors. Although both scientists and nonscientists use them, indirect metrics provide *no direct* indication of the quality of the content of the research produced by scientists.

In contradistinction, direct metrics for science-as-knowledge provide a direct measure of the quality of science content. For basic experimental science, Peter Galison identifies the metrics of directness, stability, and crucial experiments.[287] Direct metrics bring experimental reasoning another rung up the ladder of causal explanation, for example, measuring a background previously only calculated.[288] Stable measurements mean varying a feature of the experimental setup (including changes in the test substance, apparatus, arrangement, or data analysis) and obtaining the same basic result.[289] Stable measurements are desirable because each variation introduced into the experimental design makes it more difficult to postulate an alternative causal story that will satisfy all the observations; this is because the effect is nested within ever more complex loops of experimental demonstration.[290]

A final type of direct metric for basic experimental science-as-knowledge is the crucial experiment—an experiment at the "crossroads"—that yields new effects, new experimental or theoretical directions, more direct and stable measurements, or data that enable scientists to choose between competing theories.[291] It is important to note that even when scientists are involved in an entire program of related research, they almost never repeat the same experiment. Detailed case studies show that experiments are performed in series or "strings." The follow-up experiments that constitute these experiment strings" are almost always attempts to *improve* the directness and stability of

---

287 I have adopted Peter Galison's notion of directness and stability here. See Peter Galison, *How Experiments End* (Chicago: The University of Chicago Press, 1987), 259-260. Also see Mark Bodnarczuk, "Defining Metrics for Evaluating and Improving Basic Experimental Science," in *The Proceedings of the ASQC 48th Annual Quality Congress,* May 22-25, 1994, 7-9.

288 See Peter Galison, *How Experiments End* (Chicago: The University of Chicago Press, 1987), 259-260.

289 See Peter Galison, *How Experiments End* (Chicago: The University of Chicago Press, 1987), 234 ff. Latour describes roughly the same notion in terms of the "trials of strength" that various effects and substances endure at the hands of experimenters in Bruno Latour, *Science in Action* (Cambridge, MA: Harvard University Press, 1987), 74 ff.

290 See Peter Galison, *How Experiments End* (Chicago: The University of Chicago Press, 1987), 260.

291 See Ian Hacking, *Representing, and Intervening* (New York: Cambridge University Press, 1987), 249 ff.

the measurement using improved techniques for detection, data acquisition/ monitoring, or data analysis.[292]

Direct metrics for applied experimental science-as-knowledge are more tangible and include things like the development of new technological processes, development of a new technological products, reduced cost of the technologies, reduced cycle time of technologies, improved performance of technologies as evidenced in increased reliability, availability, maintainability, safety, or demonstrations of the knowledge's usefulness through commercialization. Examples of metrics for applied experimental science are legion, but I will give just a few. One example is research in wind characterization at wind turbine sites and measuring the aerodynamic phenomena in wind turbines during operation. Examples of milestones and associated metrics might include: a) performing a complete analysis of US and European aerodynamic data to define methods for gaining a 5 percent improvement in energy capture, b) developing and validating a blade-fatigue model that allows scientists to design blades with 30 percent greater lifetime than current standards, or c) developing and implementing a wind resource monitoring strategy at remote villages, and then using the data gathered as the basis for installing wind-hybrid systems over the next five years. Note that the milestones and metrics of applied science are so interrelated that they cannot be clearly separated, with the milestone being expressed as a goal or outcome and the metric expressed as the unit of measure that quantifies the goal or outcome.

Another example of applied research metrics in solar industrial technologies might be the development of new industrial applications of solar technologies and then introducing those technologies into a broad spectrum of the industrial sector. Examples of milestones and associated metrics might be things like a) negotiating and establishing cost-shared projects with three electric utilities to encourage use of solar by industrial and commercial firms, the installation of 20 solar systems, annual energy savings of 24,000 gigajoules, and the creation of 30 primary and secondary jobs, b) establishing the technical and economic feasibility of advanced ceramics production, with industry sharing 30 percent of the project costs and resources, or c) selecting three new solar advanced processes concepts for initial exploration, recording one record of invention

---

292 I discuss the details of experiment strings at Fermilab in Mark Bodnarczuk, *The Social Structure of Experimental Strings at Fermilab: A Physics and Detector Driven Model,* Fermilab: Pub-91/63, March 1990, 2 ff.

from these studies, and obtaining two expressions of interest from industrial partners for further collaborative investigations.

Metrics for Science-as-Practice

There are many stages of experimental practice that could be discussed, but I will focus on that aspect of experimental practice where the experimental design, procurements, installation, and overall configuration of the apparatus have settled down enough to actually perform the experiment. At this point, the scientist must concern him/herself with bringing the operation of the experimental apparatus into a "steady state," where all possible operational parameters of the apparatus are understood and functioning as designed. At this stage, the majority of processes that occur in the organizational infrastructure of the laboratory where an experiment is embedded exert little or no causal efficacy on the outcome of an experiment. In addition, scientists construct organizational and experimental "protective belts" around their work to protect it from all but the most devastating laboratory perturbations—usually drastic reductions in resources and funding, or in high-energy physics, the unavailability of high-energy particle beams. So how does one isolate the "vital few" activities, and support organizational interfaces that can actually affect the output (knowledge) of an experiment?

The vital few are those events that exert direct causal effects on the stability of the apparatus and computing resources because only those can affect the outcome of the experiment itself, for example, background and noise in the apparatus. Scientists control potential perturbations to the stability of the experiment by doing things like: a) visually or computationally monitoring the apparatus to ensure proper operation of components like power supplies, gases, and calibration of equipment, b) ensuring that the proper materials, targets, and chemicals are being used, c) ensuring that data rates are appropriate and that the data acquisition and software systems are functioning as designed and intended, d) ensuring that the cartographic function of experimental science is carried out in scientific notebooks and that correct data are recorded to magnetic tape, disk, or other media, and by e) ensuring that the appropriate measurement uncertainty analysis is performed on experimental results that are reported in the literature. Only these types of issues can seriously affect the production of scientific knowledge. They are a minority of laboratory activities, and they are the only activities on which scientists are interested in implementing management controls.

Our final component is indirect metrics for science-as-practice, which are an important element of the practice claims that scientists make in their proposals. This includes estimates of how long it will take to travel the experimental path, the types of experimental equipment that will be needed, the level of online and offline computing resources required, the level of engineering and technician support needed, plus time frames, milestones, human resources, and other costs. As I have described elsewhere,[293] although many scientists tend to devalue these planning and managing aspects of science, I see no fundamental reason why some of the continuous improvement methodologies developed by J. M. Juran and W. Edwards Deming cannot be tailored and applied to these practice claims.[294]

For example, scientists and laboratory managers could form cross-functional teams (composed of scientists, engineers, and technicians) that would more closely examine why scientists miss the milestones that they have defined for themselves. Following a standard quality method like storyboarding, this team could analyze the symptoms of the missed milestones, formulate theories about why they were missed, test these theories, and identify root causes. Was the problem due to a limitation that "nature" imposed on the experiment? Was it impossible to push the detector technologies any further? Was the problem an inevitable part of the pedagogic process of obtaining knowledge by actually *doing* an experiment that could not be avoided? Or was it a systemic problem where a spokesperson had no authority to make collaborators come through on their commitments? Was it a lack of planning or management on the part of the principal investigator, or the inability to stop introducing new parameters and changes into an experimental design that should have been fixed? Was the problem due to the lack of supervision of graduate students by their senior professor or laboratory manager? Problems imposed by "nature" may be unavoidable, but the practice of science would certainly be improved if most of these other problems were addressed.

Science is not, and cannot be, the only yardstick against which everything else is measured. But while science should not and cannot stand alone as the

293 See Mark Bodnarczuk, "Defining Metrics for Evaluating and Improving Basic Experimental Science," in *The Proceedings of the ASQ 48th Annual Quality Congress*, May 22-25, 1994, 12-13.

294 See W. Edwards Deming, *Out of the Crisis* (Boston, MA: The Massachusetts Inst Technology Press, 1982), and J.M. Juran, *Quality Control Handbook*, 5th Edition (New York: McGraw-Hill Professional, 1998).

final judge of what's true and meaningful, it is a powerful benchmark against which all other knowledge and experience can be and should be evaluated and calibrated. The purpose of science is not to establish truth or meaning, rather science-as-knowledge and science-as-practice are about establishing reliability, i.e., the ability to describe and predict the behavior of the universe, and the things that are in it. Science is necessary, but not sufficient to bear that burden of being the sole criterion of determining truth and meaning, without other criteria that may, or may not, be reducible to law-like, cause and effect verification or falsification.

Science enables us to measure the what, how, and when of non-unified portions, samples, of the universe and what's in it from an *outside-in* perspective. It does not measure the who and why of the universe and what's in it. Science measures the when of what happens in the universe in the *chronos* (κρόνος) sense of the word, meaning chronological time. It does not measure, and cannot measure, the when of what happens in the universe in the *kairos* (καιρός) sense of the word, i.e., an appointed time.

In addition, things look different when we see them from a distance. Things look different when we see them from an outside-in perspective (science) or an inside-out perspective (inner human experience). Perhaps one of the most poignant discussions on this point is Thomas Nagel's groundbreaking book, *What's it Like to Be a Bat?*[295] Nagel asserts that even if we could measure the properties and processes of all the neurons in a bat's brain, plus all elementary particles (quarks and leptons) that compose those neurons, and project them onto a screen (outside-in), we still would not know what it's like to *be* a bat (inside-out). Nagel's argument resonates the tendency in Western Civilization to view the universe, and the things in it, in a binary way, and then exalt one perspective over the other.

Organizational development, and management experts like Jim Collins and Jerry Porras have described this stereotypic tendency as the Tyranny of the OR, rather than the genius of the AND. I'll quote Collins and Porras at length on this point. "The 'Tyranny of the OR' pushes people to believe that things must be either A *OR* B, *but not both*. It makes such proclamations as:

- 'You can have change *OR* stability.'
- 'You can be conservative *OR* bold.'

295 Thomas Nagel, *What Is It Like to Be a Bat?* (New York: Oxford University Press, 2024).

- 'You can have low cost *OR* high quality.'
- 'You can have creative autonomy *OR* consistency and control.'
- 'You can invest for the future *OR* do well in the short-term.'
- 'You can make progress by methodical planning *OR* by opportunistic groping.'
- 'You can create wealth for your shareholders *OR* do good for the world.'
- 'You can be idealistic (values-driven) *OR* pragmatic (profit-driven).'

"Instead of being oppressed by the 'Tyranny of the OR,' highly visionary companies liberate themselves with the 'Genius of the AND'—the ability to embrace both extremes of a number of dimensions at the same time. Instead of choosing between A *OR* B, they figure out a way to have both A *AND* B."[296]

Collins and Porras go on to claim that visionary leaders who practice both-and thinking do not seek a "balance" between things like low cost and high quality; rather they relentlessly pursue *both* low cost *and* high quality at the same time. They live in the essential tension of opposites knowing that these seemingly contradictory ways of seeing actually coordinate and support each other with a synergistic kind of complementarity.[297] The word "complementarity" is another link to my future article about the ways in which the advent of quantum mechanics and the emergence of the dialogue between postmodernist, deconstructivist views of science (subjective, individual), and objective realist view science (objective, physical), made it even more difficult to press science into the service of being the ultimate adjudicator of the veracity of all human knowledge in terms of its truth and meaning.

296 This extended quote comes from James C. Collins and Jerry I Porras, *Built to Last: Successful Habits of Visionary Companies* (New York: Harper Business, 1994), 43-44 ff.

297 The view that the yin and yang principles are opposite, but mutually complementary is discussed in more detail in Alfred Huang, *The Complete I Ching* (Rochester, VT: Inner Traditions, 1998), 43 ff. Stephen Covey refers to this as Synergy—learning how to value the differences in others, see Stephen R. Covey, *The 7 Habits of Highly Effective People* (New York: A Fireside Book, 1990), 261 ff. Carl Jung discusses the essential tension and ultimate union of opposites as one of the central problems in both individual and collective (social) psychology, see C.G. Jung, "Mysterium Coniunctionis" 2nd edition in, *The Collected Works of C.G. Jung*, volume 14 (Princeton, NJ: Princeton University Press, 1989), 3 (par. 1), and 457 (par. 654).

An essential tension and complementarity is also foundational to psychotherapeutic methodologies like Cognitive Behavioral Therapy (CBT), and Dialectical Behavior Therapy (DBT), when one of the methods for improving levels of psychological health is to benchmark our inside-out, subjective, emotions, experiences, reality, and beliefs about ourselves, others, and the world around us, against a set of outside-in, objective, facts, experiences, and data in a both-and thinking, essential tension, without exalting one over the other, i.e., holding both in an essential, complementary, compensatory tension like two ends of a rubber band.

Bottom Line: Science enables us to measure the what, how, and when of non-unified portions, samples, of the universe and what's in it from an outside-in perspective. It does not measure the who and why of the universe and what's in it. Science measures the when of what happens in the universe in the *chronos* (κρόνος) sense of the word, meaning chronological time. It does not measure, and cannot measure, the when of what happens in the universe in the *kairos* (καιρός) sense of the word, i.e., an appointed time. Science can measure whether we are alive. But it cannot measure why we are alive. Science cannot measure what life's about, and whether we have a sense of the meaning of life. In this regard, I echo what the 19th-century philosopher, Soren Kierkegaard taught us about the existential now, the historical then, and the potentiality of the future, "Life can only be understood backwards; but it must be lived forwards."[298]

298 This quote comes from volume IV of the journals of the philosopher Søren Kierkegaard, written around 1843.

About the Author

Mark Bodnarczuk working in his office on the second floor of Fermilab's Wilson Hall in 1989

Mark Bodnarczuk is the Executive Director of the Breckenridge Institute®. He was on staff in the Director's Office of SLAC National Accelerator Laboratory at Stanford University from 2015 until he retired from Stanford in 2025. Mark was also in the Director's Office at Fermilab from 1985 through 1992; and in the Director's Office at the National Renewable Energy Laboratory from 1992 until 1996 when he founded the Breckenridge Institute, a management consulting firm that focuses on organizational and personal transformation and culture change. He has over forty years of experience working in the areas of basic and applied research laboratories, high-tech, pharmaceuticals, commercial construction, government, and nonprofit organizations. He also has extensive experience working in the Department of Energy National Laboratories and the National Nuclear Security Administration's nuclear weapons complex in the areas of organization development and organizational culture.

Mark has published widely on the history, social structure, and practice of high-energy physics experiments. He is editor of the anthology, *Reflections on the Fifteen Foot Bubble Chamber at Fermilab*, and the author of over twenty articles on the social structure, management and practice of high-energy physics experiments at laboratories like Fermilab, some of which appear in Lillian Hoddeson, *The Rise of the Standard Model: Particle Physics in the 1960s and 1970s* published by Cambridge University Press, and the journal *Historical Studies in the Natural Sciences* published by the University of California. While in the Fermilab Director's office, Mark was the laboratory correspondent to the *CERN Courier* where he published numerous articles on Fermilab's ongoing program in high-energy physics.

Mark is the author of five non-physics related books including *Making Invisible Bureaucracy Visible: A Guide to Assessing and Changing Organizational Culture*; *Diving In: Discovering Who You Are in the Second Half of Life*; and *Island of Excellence: 3 Powerful Strategies for Building a Culture of Creativity*. These publications explore the areas of organization development, leadership development, organizational transformation, cultural analysis, deep personal change, the Enneagram and Jungian personality typologies, and the life-long process of individuation and personal transformation in Jungian psychology, including an article entitled, *A Biblical Approach to Jung's Psychology: Experiencing the Self—the Image of God Within* published by Research Gate and Academia.edu. Mark's latest book is a memoir entitled, *Finding New Life After the Death of My Son: Grace and Forgiveness in the Age of Counterfeit Pills and Fentanyl Poisoning*.

Mark has an ABS in biblical studies from Nazarene Bible College, a BA in biblical studies from Mid-America Nazarene University, and an MA in biblical studies from Wheaton College Graduate School, and an AM in philosophy of science from the University of Chicago where he studied the conceptual development of physics from Babylonian astronomy to the Standard Model of the universe, from 1985 to 1995.

www.ingramcontent.com/pod-product-compliance
Lightning Source LLC
Chambersburg PA
CBHW050528240926
51846CB00003B/736
9781963461237